Refactoring with C++

Explore modern ways of developing maintainable and efficient applications

Dmitry Danilov

Refactoring with C++

Copyright © 2024 Packt Publishing

Group Product Manager: Kunal Sawant

Book Project Manager: Prajakta Naik

Senior Editor: Kinnari Chohan

Technical Editor: Rajdeep Chakraborthy

Copy Editor: Safis Editing

Proofreader: Kinnari Chohan

Indexer: Hemangini Bari

Production Designer: Alishon Mendonca

DevRel Marketing Coordinator: Sonia Chauhan

First published: July 2024

Production reference: 2010724

Published by Packt Publishing Ltd.

Grosvenor House

11 St Paul's Square

Birmingham

B3 1RB, UK

ISBN 978-1-83763-377-7

www.packtpub.com

This book has been a long journey, and I would like to thank my loving partner, Rina, for her unwavering support and encouragement throughout this endeavor.

I am also deeply grateful to my parents, Marina and Borys, for their continual support, inspiration, and for always encouraging me to pursue my dreams with determination.

– Dmitry Danilov

Contributors

About the author

Dmitry Danilov is an engineer and team manager with over 15 years of experience in C++. Throughout his career, he has developed network sniffers and analyzers, audio/video streaming solutions, low-level embedded applications in telecommunications, and distributed systems.

Originally from Odesa, Ukraine, Dmitry graduated with a Master's degree in Computer Engineering from Odesa National Polytechnic University. He currently resides in Tel Aviv, Israel, where he continues to advance his career in technology.

In addition to his professional career, Dmitry demonstrates his passion for knowledge sharing and engaging with the tech community through his blog (`https://ddanilov.me`) and by actively speaking at various events, including Core C++ Conference and Core C++ Group Meetup Israel.

About the reviewer

Anil Achary is a Lead Software Engineer at FactSet FinTech company with over 14 years of experience in the software industry. With expertise in C++ and real-time data processing, Anil has worked extensively in both FinTech and Telecom domains. Leading multiple engineering teams, Anil has developed high-throughput, scalable applications, optimized real-time market data systems, and integrated innovative search solutions using ChatGPT GenAI. Anil has also provided technical reviews for several books, bringing a wealth of practical knowledge to this publication.

Table of Contents

6

Utilizing a Rich Static Type System in C++   121

7

Classes, Objects, and OOP in C++   137

8

Designing and Developing APIs in C++   177

9

Code Formatting and Naming Conventions   189

10

Introduction to Static Analysis in C++   217

11

Dynamic Analysis   231

13

Modern Approach to Managing Third Parties   301

14

Version Control   321

15

Preface

In an era where higher-level languages dominate the technological landscape, C++ remains a cornerstone, driving a vast array of systems from bare-metal embedded platforms to distributed, cloud-native infrastructures. Its prowess lies in its ability to deliver performance-sensitive solutions while adeptly handling complex data structures. Over the past two decades, C++ has undergone significant evolution, continually adapting to meet the demands of modern computing.

This book serves as a comprehensive guide for those seeking to master the art of writing clean, efficient C++ code. It delves into the implementation of SOLID principles and the refactoring of legacy code using the latest features and methodologies of C++. Readers will gain a deep understanding of the language, the standard library, the extensive Boost library collection, and Microsoft's Guidelines Support Library.

Starting with the fundamentals, the book covers the core elements essential for writing clean code, with a strong emphasis on object-oriented programming in C++. It provides insights into the design principles of software testing, illustrated with examples utilizing popular unit testing frameworks like Google Test. Furthermore, the book explores the application of automated tools for both static and dynamic code analysis, featuring the powerful capabilities of Clang Tools.

By journey's end, readers will be equipped with the knowledge and skills to apply industry-approved coding practices, enabling them to craft clean, sustainable, and readable C++ code for real-world applications.

Who this book is for

This book is designed for a wide range of professionals in the C++ community. If you are a C++ engineer looking to refine your skills and write more elegant, efficient code, this book will provide you with the insights and techniques needed to elevate your programming practice. It is also an excellent resource for those tasked with refactoring and improving existing codebases, offering practical advice and strategies to make this process more manageable and effective.

Additionally, this book is an invaluable guide for tech and team leaders who aim to enhance their software development processes. Whether you are leading a small team or managing a larger development project, you will find useful tips and methodologies to make your workflows smoother and more efficient. By implementing the best practices outlined in this book, you can foster a more productive and harmonious development environment, ultimately leading to higher-quality software and more successful projects.

What this book covers

Chapter 1, Coding Standards in C++, explores the world of clean code and its crucial role in successful software projects. We discuss technical debt and how poor-quality code contributes to its accumulation. The chapter also covers the importance of code formatting and documentation, emphasizing their role in maintaining a manageable and effective codebase. We introduce common conventions and best practices used in the C++ community, highlighting the necessity of clean code and proper documentation for any project.

Chapter 2, Main Software Development Principles, covers key software design principles for creating well-structured and maintainable code. We discuss the SOLID principles—Single Responsibility, Open-Closed, Liskov Substitution, Interface Segregation, and Dependency Inversion—which help developers write code that is easy to understand, test, and modify. We also highlight the importance of levels of abstraction, the concepts of side effects and mutability, and their impact on software quality. By applying these principles, developers can create more robust, reliable, and scalable software.

Chapter 3, Causes of Bad Code, identifies the key factors leading to subpar code in C++. These include the pressure to deliver quickly, the flexibility of C++ allowing multiple solutions to the same problem, personal coding styles, and a lack of knowledge of modern C++ features. Understanding these causes helps developers avoid common pitfalls and improve existing codebases effectively.

Chapter 4, Identifying Ideal Candidates for Rewriting - Patterns and Anti-Patterns, focuses on identifying ideal candidates for refactoring in C++ projects. We will explore factors that make code segments suitable for refactoring, such as technical debt, complexity, and poor readability. We will also discuss common patterns and anti-patterns, providing guidelines and techniques for improving code structure, readability, and maintainability without altering its behavior. This chapter aims to equip developers with the knowledge to enhance the quality and robustness of their C++ codebases effectively.

Chapter 5, The Significance of Naming, highlights the crucial role of naming conventions in C++ programming. Proper names for variables, functions, and classes enhance code readability and maintainability. We discuss best practices for naming, the impact of poor naming on code efficiency, and the importance of consistent coding conventions. By understanding and applying these principles, you will write clearer and more effective code.

Chapter 6, Utilizing a Rich Static Type System in C++, explores the powerful static type system in C++, emphasizing its role in writing robust, efficient, and maintainable code. We discuss advanced techniques like using the `<chrono>` library for time durations, `not_null` wrappers, and `std::optional` for safer pointer handling. Additionally, we look at external libraries like Boost for enhancing type safety. Through real-world examples, we demonstrate how to leverage these tools to harness the full potential of C++'s type system, resulting in more expressive and error-resistant code.

Chapter 7, Classes, Objects, and OOP in C++, focuses on advanced topics in classes, objects, and object-oriented programming (OOP) in C++. We cover class design, method implementation, inheritance, and template usage. Key topics include optimizing class encapsulation, advanced method practices, evaluating inheritance versus composition, and sophisticated template techniques. Practical examples illustrate these concepts, helping you create robust and scalable software architectures.

Chapter 8, Designing and Developing APIs in C++, explores the principles and practices for designing maintainable APIs in C++. We discuss the importance of clarity, consistency, and extensibility in API design. Through concrete examples, we illustrate best practices that help create intuitive, easy-to-use, and robust APIs. By applying these principles, you will develop APIs that meet user needs and remain adaptable over time, ensuring the longevity and success of your software libraries.

Chapter 9, Code Formatting and Naming Conventions, explore the critical role of code formatting and naming conventions in creating robust and maintainable software. While these topics may seem minor, they greatly enhance code readability, simplify maintenance, and foster effective team collaboration, especially in complex languages like C++. We delve into the importance of code formatting and provide practical knowledge on using tools like Clang-Format and editor-specific plugins to implement consistent formatting. By the end of this chapter, you'll understand the significance of these practices and how to apply them effectively in your C++ projects.

Chapter 10, Introduction to Static Analysis in C++, discusses the crucial role of static analysis in ensuring code quality and reliability in C++ development. We discuss how static analysis identifies bugs quickly and cost-effectively, making it a key component of software quality assurance. We delve into popular tools like Clang-Tidy, PVS-Studio, and SonarQube, and guide integrating static analysis into your development workflow.

Chapter 11, Dynamic Analysis, explores dynamic code analysis in C++, focusing on tools that scrutinize program behavior during execution to detect issues like memory leaks, race conditions, and runtime errors. We cover compiler-based sanitizers such as Address Sanitizer (ASan), Thread Sanitizer (TSan), and Undefined Behavior Sanitizer (UBSan), along with Valgrind for thorough memory debugging. By understanding and integrating these tools into your development workflow, you can ensure cleaner, more efficient, and reliable C++ code.

Chapter 12, Testing, emphasizes the crucial role of software testing in ensuring quality, reliability, and maintainability. We cover various testing methodologies, starting with unit testing to validate individual components, followed by integration testing to examine the interaction between integrated units. We then move to system testing for a comprehensive assessment of the entire software system and conclude with acceptance testing to ensure the software meets end-user requirements. By understanding these methodologies, you will grasp how testing underpins the development of robust and user-centric software.

Chapter 13, Modern Approach to Managing Third Parties, addresses the critical role of third-party libraries in C++ development. We explore the basics of third-party library management, including the impact of static versus dynamic compilation on deployment. Given C++'s lack of a standardized library ecosystem, we examine tools like vcpkg and Conan to understand their advantages in integrating and managing libraries. Additionally, we discuss the use of Docker for creating consistent and reproducible development environments. By the end of this chapter, you'll be equipped to choose and manage third-party libraries effectively, enhancing your development workflow and software quality.

Chapter 14, Version Control, highlights the importance of maintaining a clean commit history in software development. We discuss best practices for clear and purposeful commit messages and introduce tools like Git, Conventional Commit Specification, and commit linting. By following these principles, developers can enhance communication, collaboration, and project maintainability.

Chapter 15, Code Review, explores the critical role of code review in ensuring robust and maintainable C++ code. While automated tools and methodologies provide significant benefits, they are not foolproof. Code reviews, conducted by human reviewers, help catch errors that automated processes may miss and ensure adherence to standards. We discuss strategies and practical guidelines for effective code reviews, emphasizing their role in preventing bugs, enhancing code quality, and fostering a collaborative culture of learning and accountability.

To get the most out of this book

To get the most out of this book, you should have a solid understanding of the basics of C++. Familiarity with build systems such as Make and CMake will be beneficial. Additionally, having basic knowledge of Docker and terminal skills can enhance your learning experience, although these are optional.

If you are using the digital version of this book, we advise you to type the code yourself or access the code from the book's GitHub repository (a link is available in the next section). Doing so will help you avoid any potential errors related to the copying and pasting of code.

Download the example code files

You can download the example code files for this book from GitHub at `https://github.com/PacktPublishing/Refactoring-with-C-`. If there's an update to the code, it will be updated in the GitHub repository.

We also have other code bundles from our rich catalog of books and videos available at `https://github.com/PacktPublishing/`. Check them out!

Conventions used

There are a number of text conventions used throughout this book.

`Code in text`: Indicates code words in text, database table names, folder names, filenames, file extensions, pathnames, dummy URLs, user input, and Twitter handles. Here is an example: "Mount the downloaded `WebStorm-10*.dmg` disk image file as another disk in your system."

A block of code is set as follows:

```
html, body, #map {
  height: 100%;
  margin: 0;
  padding: 0
}
```

When we wish to draw your attention to a particular part of a code block, the relevant lines or items are set in bold:

```
[default]
exten => s,1,Dial(Zap/1|30)
exten => s,2,Voicemail(u100)
exten => s,102,Voicemail(b100)
exten => i,1,Voicemail(s0)
```

Any command-line input or output is written as follows:

```
$ mkdir css
$ cd css
```

Bold: Indicates a new term, an important word, or words that you see onscreen. For instance, words in menus or dialog boxes appear in **bold**. Here is an example: "Select **System info** from the **Administration** panel."

> **Tips or important notes**
> Appear like this.

Get in touch

Feedback from our readers is always welcome.

General feedback: If you have questions about any aspect of this book, email us at customercare@ packtpub.com and mention the book title in the subject of your message.

Errata: Although we have taken every care to ensure the accuracy of our content, mistakes do happen. If you have found a mistake in this book, we would be grateful if you would report this to us. Please visit www.packtpub.com/support/errata and fill in the form.

Piracy: If you come across any illegal copies of our works in any form on the internet, we would be grateful if you would provide us with the location address or website name. Please contact us at copyright@packt.com with a link to the material.

If you are interested in becoming an author: If there is a topic that you have expertise in and you are interested in either writing or contributing to a book, please visit authors.packtpub.com.

Share Your Thoughts

Once you've read *Refactoring with C++*, we'd love to hear your thoughts! Scan the QR code below to go straight to the Amazon review page for this book and share your feedback.

https://packt.link/r/1837633770

Your review is important to us and the tech community and will help us make sure we're delivering excellent quality content.

Download a free PDF copy of this book

Thanks for purchasing this book!

Do you like to read on the go but are unable to carry your print books everywhere?

Is your eBook purchase not compatible with the device of your choice?

Don't worry, now with every Packt book you get a DRM-free PDF version of that book at no cost.

Read anywhere, any place, on any device. Search, copy, and paste code from your favorite technical books directly into your application.

The perks don't stop there, you can get exclusive access to discounts, newsletters, and great free content in your inbox daily

Follow these simple steps to get the benefits:

1. Scan the QR code or visit the link below

https://packt.link/free-ebook/9781837633777

2. Submit your proof of purchase
3. That's it! We'll send your free PDF and other benefits to your email directly

Coding Standards in C++

In this chapter, we will delve into the world of clean code and examine its crucial role in successful software projects. We will discuss the concept of technical debt and how poor-quality code can contribute to its accumulation. Additionally, we will explore the often undervalued practices of code formatting and documentation, which are crucial for maintaining a manageable and effective code base. Through this chapter, we will understand that clean code is not just nice to have but a necessity for any project.

We will discuss the importance of coding standards and introduce common conventions and best practices used in the C++ community. By the end of this chapter, you will better understand what clean code is, why it is essential, and why documenting code is crucial.

The difference between good code and bad code

There is no strict definition of good or clean code. Moreover, no automatic tool can measure the quality of code. There are linters, code checkers, and other analyzers that can help to make code better. These tools are very valuable and highly recommended, but not sufficient. Artificial Intelligence may take over and develop code for us, but in the end, its measurement of code quality will be based on our human ideas of good code.

Programming languages were initially developed to provide an interface between machines and developers; however, with the growth of software products' complexity, it becomes clear that nowadays, it is mostly a way of communicating ideas and intentions between developers. It is a well-known fact that developers spend ten times more time reading the code than writing it. It means that to be efficient, we must do our best to make the reading easier. The most successful way to make this process efficient is to make the code predictable or, even better, boring. By boring, I mean code that the reader looks at and knows what to expect from it regarding functionality, performance, and side effects. The following example illustrates an interface for a class retrieving objects from a database:

```cpp
class Database {
public:
  template<typename T>
  std::optional<T> get(const Id& id) const;
```

```
template<typename T>
std::optional<T> get(const std::string& name) const;
};
```

It supports two lookup modes, by `id` and by `name`; it is not supposed to change the internal state of the `Database` object because of the `const` modifier; and it can only do read operations against the database instance. It is boring, but it meets basic expectations. Imagine how surprising it can be to find out during critical bug investigation that on each read operation, it does an update operation and sometimes a delete operation?

As readers can see, a few key factors can distinguish good code and bad code. Good code is typically well written, easy to read, and efficient. It follows the conventions and standards of the C++ language and is organized in a logical and consistent manner. Good code is also well documented, often with clear comments explaining the purpose and function of things that are not obvious from reading only the code.

Why coding standards are important

Coding standards are important for several reasons. First and foremost, they help to minimize *technical debt*. Technical debt, also known as "code debt," is a metaphor that describes the cost of maintaining and modifying code that is not well designed or well written. Just as financial debt incurs interest and requires ongoing payments, technical debt incurs additional costs in the form of the time and effort needed to maintain and modify poorly designed code.

Technical debt can accumulate in many ways, such as through hacky or quick-fix solutions to problems or by ignoring best practices or coding standards. As technical debt accumulates, it can become increasingly difficult and time-consuming to modify and maintain the code, which can negatively impact the efficiency and effectiveness of a development team.

In order to manage technical debt carefully, it is essential to try to avoid accumulating too much of it, as it can become a significant burden on a development team. Strategies for managing technical debt include regularly refactoring code to improve its design and maintainability, following best practices and coding standards, and actively seeking out opportunities to improve the quality of the code. Overall, managing technical debt is an important aspect of good code design and development and can help to ensure that code is efficient, reliable, and easy to work with.

Coding standards help to ensure the quality and consistency of code. By establishing a set of guidelines and conventions for writing code, coding standards help to ensure that code is well written, easy to read, and easy to understand. This makes it easier for others to maintain and update the code and helps to prevent errors and bugs.

In addition, they help to improve the efficiency of code. By following established conventions and best practices, programmers can write code that is more efficient and performs better. This can save time and resources and help ensure that code is scalable and can handle large amounts of data and traffic.

Furthermore, coding standards promote collaboration and teamwork among programmers. By establishing a common set of guidelines and conventions, coding standards make it easier for teams of programmers to work together on a project. This allows better communication and coordination and helps to ensure that everyone is on the same page and working towards the same goals.

Coding standards often promote the interoperability and portability of code. By following a standardized set of conventions, code written by one programmer can be easily understood and used by another programmer. This allows code to be more easily integrated into larger projects and helps to ensure that it can be used on a variety of different platforms and operating systems.

The C++ programming language is probably one of the richest languages in terms of features. It started as *C with classes*, providing object-oriented support with high performance and almost complete compatibility with C; later on, template metaprogramming was introduced, and Stepanov and Lee developed the Standard Template Library, nowadays known as the C++ Standard Library. Modern C++ (C++11 and newer versions) provides extensive support for multiple programming paradigms, including procedural, object-oriented, generic, and functional programming. It offers features such as lambda expressions, range-based `for` loops, smart pointers, and type inference that enable functional programming techniques. Additionally, C++ provides support for object-oriented programming concepts such as inheritance, encapsulation, and polymorphism. It also offers template metaprogramming, which enables generic programming and allows compile-time optimizations. Furthermore, C++ provides concurrency support with features such as threads, atomic types, and futures, making it easier to write concurrent and parallel code. This flexibility is key to the strength of the language but often leads to problems with maintainability.

A developer has to understand the concepts of the paradigms we've mentioned, how to use them together, and how they eventually affect the performance of the code. This is when coding standards can help to explain the complexity of the code base.

All these factors make coding guidelines the bare minimum that a modern C++ project should have to attain a quality standard.

Code convention

Contrary to languages such as Python, Go, Java, and many others, C++ does not have a common code convention.

There are several popular coding conventions for the C++ programming language. Here are some common conventions that are widely followed:

- **Naming conventions**: Naming conventions specify how to name variables, functions, classes, and other code elements in a descriptive manner. For example, variables may be named using lowercase letters, with words separated by underscores, such as `total_cost` or `customer_name`. Class variables often have prefixes or suffixes to distinguish them from other variables, such as `m_user_count` or `user_count_`. Functions may be named using *camelCase*, with the first letter of each word (apart from the first word) capitalized, such as `calculateTotalCost` or `getCustomerName`. Classes may be named using *PascalCase*, with the first letter of each word capitalized, such as `Customer` or `Invoice`.

- **Commenting**: Commenting conventions specify how to write and format comments in code. Comments are used to provide explanations and documentation for code and should be clear and concise. It is often recommended to use inline comments to explain specific lines of code, as well as block comments to provide an overview of a code block or function.

- **Formatting**: Formatting conventions specify how to format code for readability and consistency. This may include conventions for indentation, spacing, line breaks, and other elements of code layout. For example, it is common to indent code blocks, such as those inside a `for` loop or `if` statement, to visually indicate the structure of the code. Formatting policy often covers asterisk (`*`) and ampersand (`&`) alignment in pointers and references (e.g., `int* ptr` versus `int *ptr` or `Socket &socket` versus `Socket& socket`), curly braces position (same line, next line, or context dependent). This book covers aspects of automated formatting in *Chapter 13*.

- **Coding style:** Coding style conventions specify how to write and structure code for clarity and readability. This may include conventions for variable declarations, control flow, and other elements of code structure, such as how variables are passed to functions (by value, reference, or pointer) and the usage of specific language features, such as exceptions and `goto` operators.

It is important to note that there may be variations in coding conventions between different organizations and teams. It is important to follow the conventions that your team or organization establishes or to define your own conventions if none are specified.

It can be tedious to develop a coding convention; some companies prefer to use an existing one and adapt it to their needs. In the C++ programming language, there are several popular code standards that are widely followed by developers. These standards aim to improve the readability, maintainability, and overall quality of C++ code.

One common code standard for C++ is the C++ Core Guidelines (`https://isocpp.github.io/CppCoreGuidelines/CppCoreGuidelines`), which were developed by Bjarne Stroustrup, the creator of C++, and a group of experts from industry and academia. The guidelines cover a wide range of topics, including naming conventions, commenting, formatting, and coding style.

Another popular code standard for C++ is the Google C++ Style Guide (`https://google.github.io/styleguide/cppguide.html`), which is used by many software companies, including Google. The guide provides guidelines for naming conventions, commenting, formatting, and coding style, as well as recommendations for using specific C++ features and libraries.

In addition to these widely followed standards, there are also many other code standards that have been developed by individual organizations and teams, for example, LLVM Coding Standards, WebKit, and Mozilla's style guides.

If a project conforms with a specific code convention, it is easier to read it, and as a bonus, the code base becomes more *grepable*. Consider needing to find the places where a variable called `request_id` is assigned. It can be easily achieved via the `grep` utility:

```
$ grep -rn "request_id = " .
./RequestHandler.cpp:25: request_id = new_request_id;
./RequestHandler.cpp:122: request_id = request.getId();
```

Code reviewers used to spend hours catching and commenting on inconsistencies with code format during peer review. Luckily, today we have tools such as Clang-Tidy and Clang-Format that allow us to ensure the consistency of the code format automatically via code editors and **continuous integration** (**CI**). We will dive deeper into their configuration later in this book in *Chapter 10*.

Language features limitations

C++ is a powerful language; as we know, great power comes with great responsibility. It is not easy for engineers, especially those who have not spent decades writing C++ code, to grasp the complexity of the language. As a result, some companies decide to limit the features used in their projects. The limitations may include a ban for multiple inheritance, usage of exceptions, and minimal usage of macros, templates, and specific third-party libraries. Additionally, the regulations may come from the use of legacy libraries. For example, if most of the code does not support C++ exceptions, it might be a bad idea to add them to new pieces of code without a prior understanding of the outcome.

General guidelines

It is always a good idea to have general guidelines for a project. The guidelines often cover the preferred way of working on the project:

- Usage of raw pointers, if allowed
- How values are returned from getters and provided to setters (by value or reference)
- Use of code comments:
  - Are comments allowed in general?
  - Usage of strategic and tactical comments
  - Comment style: free, Doxygen, and so on

Coding standards are necessary to help ensure code quality, consistency, interoperability, portability, efficiency, and collaboration. By following coding standards, programmers can write better code that is easier to understand, maintain, and use and works better and more efficiently.

Readability, efficiency, maintainability, and usability

Readability, efficiency, maintainability, and usability are all critical factors to consider when writing code.

Readability

Readability refers to the ease with which a human reader can understand a piece of code. Well-written code is easy to read, with clear and concise statements that are organized in a logical and consistent manner. This becomes very important if we consider that developers spend ten times more time reading code than writing it.

Let's take a look at the following piece of code:

```
class Employee {
public:
  std::string get_name();
  std::string surname();
  uint64_t getId() const;
};
```

This example is an exaggerated example of code not following any code convention. A developer using the Employee class can understand that all three methods are getters. However, the differences in the names make the user spend more time understanding the code or trying to understand the reasoning behind the names. Do the methods have different names because programmers did not care about the uniformity of the class? Or because, for example, methods without the get prefix are simple getters, and ones that contain the get prefix fetch the data from a file or a database?

Additionally, do the methods without const change the object's state (via caching, for example), or is it a mistake? Do you see how many questions can be raised? They can be answered only when a developer jumps into the corresponding implementation, which means time is wasted. Making the code look uniform across the code base helps developers to understand the meanings and complications of classes, methods, and functions by looking at their declarations in the header files or even via autocompletion in modern code editors.

Efficiency

Efficiency refers to the ability of a piece of code to perform its intended tasks in an efficient manner. Efficient code uses few resources, such as time and memory, to accomplish a task, and is able to handle large amounts of data and traffic without slowing down or crashing. By improving the efficiency of code, programmers can save time and resources and can ensure that their code is scalable and can handle the demands of a growing user base.

There are surefire ways to optimize C++ code, such as passing read-only parameters by constant reference to avoid unnecessary copying, or using the character-overloaded version of `std::string::find` when looking for a single character to avoid string a creation:

```
my_string.find('A');
my_string.find("A");
```

However, the more systematic way to achieve and maintain code efficiency is to follow the Pareto principle. This principle, when applied to software engineering, says that roughly 20% of the code does 80% of the work. For example, usually, there is no need to optimize the code parsing config files on startup of a background daemon because it happens only once during the program's lifetime. However, it might be important to avoid copying large data structures in the main flow. The optimal way to improve the efficiency includes picking this 20% of performance-critical code and adding benchmarks for it. The benchmarks are expected to run as part of the CI process to make sure that no degradation is introduced.

Additionally, end-to-end testing can measure the overall performance of the application. This book discusses the best practices of writing unit tests and end-to-end tests in *Chapter 13*. It is important to note that the automated tools cannot replace an engineer doing code reviews for the new code, mainly because there is no tool that can find the 20% of the code that does 80% of the work.

Maintainability

Maintainability refers to the ease with which a piece of code can be updated and modified over time. Well-written code is easy to maintain, with clear and well-documented code that is organized in a logical and consistent manner. By improving the maintainability of code, programmers can make it easier for others to update and modify their code and can ensure that their code remains relevant and useful over time. Ideally, while developing new components, developers should think about the current problems the code is solving and the future usage and extension of the code. For example, while developing support for a data provider, it might be useful to ask whether the provider is going to be the only one supported. If not, it might be helpful to think about the standard features of data providers and extract them in an abstract base class. Here's an example:

```
class BaseDataProvider {
public:
  BaseDataProvider() = default;
```

```
   BaseDataProvider(const BaseDataProvider&) = delete;
   BaseDataProvider(BaseDataProvider&&) = default;
   BaseDataProvider& operator = (const BaseDataProvider&) =
      delete;
   BaseDataProvider& operator = (BaseDataProvider&&) =
      default;
   virtual ~BaseDataProvider() = default;

   virtual Data getData() const = 0;
};

class NetworkDataProvider : public BaseDataProvider {
public:
   NetworkDataProvider(const Endpoint& endpoint);
   Data getData() const override;
};

class FileDataProvider : public BaseDataProvider {
public:
   FileDataProvider(const std::string& filename);

   Data getData() const override;
};
```

In this example, the `DataProvider` class is an abstract base class that defines the interface for providing data. The `NetworkDataProvider` and `FileDataProvider` classes are derived from `DataProvider` and override the `getData` virtual function to provide the specific implementation for reading data from a file or a network endpoint, respectively. This design makes it easy to add new data sources by simply creating a new class derived from `DataProvider` and providing the appropriate implementation for the `getData` virtual function.

It is clear from the example that the base interface may include not only the functionality but also the copy-move policy of the object. Later, the user code can receive the data provider(s) with reference to the base class and be agnostic to the type of provider, as shown in the following snippet:

```
class DataParser {
public:
   DataParser(const BaseDataProvider& provider);
   void parse();
};
```

Additionally, this inheritance can be used for mocking data providers while creating unit tests for `DataParser`. Unit tests are covered in detail in *Chapter 13*.

On a side note, it is crucial not to make code overcomplicated, or to be ready for any change. Otherwise, the need to make everything extendable may lead to monsters such as the following snippet:

```cpp
#define BASE_CLASS(TYPE)                     \
  template <typename T>                      \
  class TYPE {                               \
  public:                                    \
    T value;                                 \
    TYPE(T val) : value(val) {}              \
  };

#define DERIVED_CLASS(TYPE, BASE)            \
  template <typename T>                      \
  class TYPE : public BASE<T> {              \
  public:                                    \
    TYPE(T val) : BASE<T>(val) {}            \
    T getValue() { return value; } \
  };

BASE_CLASS(Base);
DERIVED_CLASS(Derived, Base);

int main() {
  Derived<int> obj(5);
  std::cout << obj.getValue() << std::endl;
  return 0;
}
```

This class hierarchy is unnecessarily complicated because it uses almost every C++ feature: inheritance, templates, and macros. While using inheritance with templates is a common practice, macros are seen as an anti-pattern nowadays. In this example, the `Derived` class adds very little additional functionality compared to the `Base` class, and it would be more straightforward to simply add the `getValue` method directly to the `Base` class. Using inheritance and templates can be useful in certain situations, but it's important to use them appropriately and not overuse them. Macros can be particularly difficult to understand and maintain because they are expanded by the preprocessor before the code is compiled, so it can be hard to see what the actual code looks like. It's generally better to use functions or template functions instead of macros whenever possible.

If the probability of extension is low, keeping its structure simple and close to basic needs is better. How do you decide what approach to take? Well, calm consideration and code review is the way to find out.

Usability

Usability refers to the ease with which a piece of code can be used by others. Well-written code is easy to use, with clear and intuitive interfaces and documentation that make it easy for others to understand and use the code. By improving the usability of code, programmers can make their code more accessible and useful to others and can ensure that their code is widely adopted and used.

Overall, readability, efficiency, maintainability, and usability are all important factors to consider when writing code. By improving these factors, programmers can write better code that is easier to understand, maintain, and use.

Summary

In this chapter, you learned about the concept of good and bad code. Good code is well written, efficient, and easy to understand and maintain. It follows coding standards and best practices and is less prone to errors. On the other hand, bad code is poorly written, inefficient, and difficult to understand and maintain.

The chapter also introduced the concept of technical debt, which refers to the accumulation of poor-quality code that needs to be refactored or rewritten. Technical debt can be costly and time-consuming to fix and can hinder the development of new features or functionality.

The importance of code standards was also emphasized in the chapter. Code standards are guidelines or rules that dictate how code should be written, formatted, and structured. Adhering to code standards helps to ensure that code is consistent, easy to understand, and maintainable. It also makes it easier for multiple developers to work on the same code base and helps to prevent errors and bugs.

Overall, the chapter emphasized the importance of writing good quality code and adhering to code standards in order to avoid technical debt and ensure the long-term success and maintainability of a software project.

In the next chapter, we will dive into the world of software design principles. Specifically, we will focus on the SOLID principles, a set of guidelines that aim to improve the design of software systems by making them more maintainable, flexible, and scalable. Each of the principles will be explained in detail in the next chapter, along with examples of how they can be applied to real-world software development scenarios.

## 2

Main Software Development Principles

In this chapter, we will explore the main software design principles that are used to create well-structured and maintainable code. One of the most important principles is SOLID, which stands for Single Responsibility Principle, Open-Closed Principle, Liskov Substitution Principle, Interface Segregation Principle, and Dependency Inversion Principle. These principles are designed to help developers create code that is easy to understand, test, and modify. We will also discuss the importance of levels of abstraction, which is the practice of breaking down complex systems into smaller, more manageable parts. Additionally, we will explore the concepts of side effects and mutability and how they can affect the overall quality of software. By understanding and applying these principles, developers can create software that is more robust, reliable, and scalable.

SOLID

SOLID is a set of principles that were first introduced by Robert C. Martin in his book *Agile Software Development, Principles, Patterns, and Practices*, in 2000. Robert C. Martin, also known as Uncle Bob, is a software engineer, author, and speaker. He is considered one of the most influential figures in the software development industry, known for his work on the SOLID principles and his contributions to the field of object-oriented programming. Martin has been a software developer for more than 40 years and has worked on a wide variety of projects, from small systems to large enterprise systems. He is also a well-known speaker and has given presentations on software development at many conferences and events around the world. He is an advocate of agile methodologies, and he has been influential in the development of the Agile Manifesto. The SOLID principles were developed as a way to help developers create more maintainable and scalable code by promoting good design practices. The principles were based on Martin's experience as a software developer and his observation that many software projects suffer from poor design, which makes them difficult to understand, change, and maintain over time.

The SOLID principles are intended to be a guide for object-oriented software design, and they are based on the idea that software should be easy to understand, change, and extend over time. The principles are meant to be applied in conjunction with other software development practices, such as test-driven development and continuous integration. By following SOLID principles, developers can create code that is more robust, less prone to bugs, and easier to maintain over time.

The Single Responsibility Principle

The **Single Responsibility Principle (SRP)** is one of the five SOLID principles of object-oriented software design. It states that a class should have only one reason to change, meaning that a class should have only one responsibility. This principle is intended to promote code that is easy to understand, change, and test.

The idea behind the SRP is that a class should have a single, well-defined purpose. This makes it easier to understand the class's behavior and makes it less likely that changes to the class will have unintended consequences. When a class has only one responsibility, it is also less likely to have bugs, and it is easier to write automated tests for it.

Applying the SRP can be a useful way to improve the design of a software system by making it more modular and easier to understand. By following this principle, a developer can create classes that are small, focused, and easy to reason about. This makes it easier to maintain and improve the software over time.

Let us look at a messaging system that supports multiple message types sent over the network. The system has a `Message` class that receives sender and receiver IDs and raw data to be sent. Additionally, it supports saving messages to the disk and sending itself via the `send` method:

```
class Message {
public:
  Message(SenderId sender_id, ReceiverId receiver_id,
          const RawData& data)
    : sender_id_{sender_id},
      receiver_id_{receiver_id}, raw_data_{data} {}

  SenderId sender_id() const { return sender_id_; }
  ReceiverId receiver_id() const { return receiver_id_; }

  void save(const std::string& file_path) const {
    // serializes a message to raw bytes and saves
    // to file system
  }

  std::string serialize() const {
    // serializes to JSON
```

```
    return {"JSON"};
  }

  void send() const {
    auto sender = Communication::get_instance();
    sender.send(sender_id_, receiver_id_, serialize());
  }

private:
  SenderId sender_id_;
  ReceiverId receiver_id_;
  RawData raw_data_;
};
```

The Message class is responsible for multiple concerns, such as saving messages from/to the filesystem, serializing data, sending messages, and holding the sender and receiver IDs and raw data. It would be better to separate these responsibilities into different classes or modules.

The Message class is only responsible for storing the data and serializing it to JSON format:

```
class Message {
public:
  Message(SenderId sender_id, ReceiverId receiver_id,
          const RawData& data)
    : sender_id_{sender_id},
      receiver_id_{receiver_id}, raw_data_{data} {}

  SenderId sender_id() const { return sender_id_; }
  ReceiverId receiver_id() const { return receiver_id_; }

  std::string serialize() const {
    // serializes to JSON
    return {"JSON"};
  }

private:
  SenderId sender_id_;
  ReceiverId receiver_id_;
  RawData raw_data_;
};
```

The `save` method can be extracted to a separate `MessageSaver` class, having a single responsibility:

```
class MessageSaver {
public:
  MessageSaver(const std::string& target_directory);
  void save(const Message& message) const;
};
```

And the `send` method is implemented in a dedicated `MessageSender` class. All three classes have a single and clear responsibility, and any further changes in any of them would not affect the others. This approach allows isolating the changes in the code base. It becomes crucial in a complex system requiring long compilation.

In summary, the SRP states that a class should have only one reason to change, meaning that a class should have only one responsibility. This principle is intended to promote code that is easy to understand, change, and test, and it helps in creating a more modular, maintainable, and scalable code base. By following this principle, developers can create classes that are small, focused, and easy to reason about.

Other applications of the SRP

The SRP can be applied not only to classes but also to larger components, such as applications. At the architecture level, the SRP is often implemented as microservices architecture. The idea of microservices is to build a software system as a collection of small, independent services that communicate with each other over a network rather than building it as a monolithic application. Each microservice is *responsible for a specific business capability and can be developed, deployed, and scaled independently from the other services.* This allows for greater flexibility, scalability, and ease of maintenance, as changes to one service do not affect the entire system. Microservices also enable a more agile development process, as teams can work on different services in parallel, and also allows for a more fine-grained approach to security, monitoring, and testing, as each service can be handled individually.

The Open-Closed Principle

The Open-Closed principle states that a module or class should be open for extension but closed for modification. In other words, it should be possible to add new functionality to a module or class without modifying its existing code. This principle helps to promote software maintainability and flexibility. An example of this principle in C++ is the use of inheritance and polymorphism. A base class can be written with the ability to be extended by derived classes, allowing for new functionality to be added without modifying the base class. Another example is using interfaces or abstract classes to define a contract for a set of related classes, allowing new classes to be added that conform to the contract without modifying existing code.

The Open-closed Principle can be used to improve our message-sending components. The current version supports only one message type. If we want to add more data, we need to change the `Message` class: add fields, hold a message type as an additional variable, and not to mention serialization based on this variable. In order to avoid changes in existing code, let us rewrite the `Message` class to be purely virtual, providing the `serialize` method:

```
class Message {
public:
  Message(SenderId sender_id, ReceiverId receiver_id)
    : sender_id_{sender_id}, receiver_id_{receiver_id} {}

  SenderId sender_id() const { return sender_id_; }
  ReceiverId receiver_id() const { return receiver_id_; }
  virtual std::string serialize() const = 0;

private:
  SenderId sender_id_;
  ReceiverId receiver_id_;
};
```

Now, let us assume that we need to add another two message types: a "start" message supporting start delay (often done for debugging purposes) and a "stop" message supporting stop delay (can be used for scheduling); they can be implemented as follows:

```
class StartMessage : public Message {
public:
  StartMessage(SenderId sender_id, ReceiverId receiver_id,
               std::chrono::milliseconds start_delay)
    : Message{sender_id, receiver_id},
      start_delay_{start_delay} {}

  std::string serialize() const override {
    return {"naive serialization to JSON"};
  }

private:
  const std::chrono::milliseconds start_delay_;
};

class StopMessage : public Message {
public:
  StopMessage(SenderId sender_id, ReceiverId receiver_id,
              std::chrono::milliseconds stop_delay)
    : Message{sender_id, receiver_id},
```

```
          stop_delay_{stop_delay} {}

   std::string serialize() const override {
     return {"naive serialization to JSON"};
   }

private:
  const std::chrono::milliseconds stop_delay_;
};
```

Note that none of the implementations requires changes in other classes, and each of them provides its own version of the `serialize` method. The `MessageSender` and `MessageSaver` classes do not need additional adjustments to support the new class hierarchy of messages. However, we are going to change them too. The main reason is to make them extendable without requiring changes. For example, a message can be saved not only to the filesystem but also to remote storage. In this case, `MessageSaver` becomes purely virtual:

```
class MessageSaver {
public:
  virtual void save(const Message& message) const = 0;
};
```

The implementation responsible for saving to the filesystem is a class derived from `MessageSaver`:

```
class FilesystemMessageSaver : public MessageSaver {
public:
  FilesystemMessageSaver(const std::string&
    target_directory);
  void save(const Message& message) const override;
};
```

And the remote storage saver is another class in the hierarchy:

```
class RemoteMessageSaver : public MessageSaver {
public:
    RemoteMessageSaver(const std::string&
      remote_storage_address);
    void save(const Message& message) const override;
};
```

The Liskov Substitution Principle

The **Liskov Substitution Principle (LSP)** is a fundamental principle in object-oriented programming that states that objects of a superclass should be able to be replaced with objects of a subclass without affecting the correctness of the program. This principle is also known as the Liskov principle, named after Barbara Liskov, who first formulated it. The LSP is based on the idea of inheritance and polymorphism, where a subclass can inherit the properties and methods of its parent class and can be used interchangeably with it.

In order to follow the LSP, subclasses must be "behaviorally compatible" with their parent class. This means that they should have the same method signatures and follow the same contracts, such as input and output types and ranges. Additionally, the behavior of a method in a subclass should not violate any of the contracts established in the parent class.

Let's consider a new `Message` type, `InternalMessage`, which does not support the `serialize` method. One might be tempted to implement it in the following way:

```
class InternalMessage : public Message {
public:
    InternalMessage(SenderId sender_id, ReceiverId
      receiver_id)
        : Message{sender_id, receiver_id} {}

    std::string serialize() const override {
        throw std::runtime_error{"InternalMessage can't be
          serialized!"};
    }
};
```

In the preceding code, `InternalMessage` is a subtype of `Message` but cannot be serialized, throwing an exception instead. This design is problematic for a few reasons:

- **It breaks the Liskov Substitution Principle**: As per the LSP, if `InternalMessage` is a subtype of `Message`, then we should be able to use `InternalMessage` wherever `Message` is expected without affecting the correctness of the program. By throwing an exception in the `serialize` method, we are breaking this principle.

- **The caller must handle exceptions**: The caller of `serialize` must handle exceptions, which might not have been necessary when dealing with other `Message` types. This introduces additional complexity and the potential for errors in the caller code.

- **Program crashes**: If the exception is not properly handled, it could lead to the program crashing, which is certainly not a desirable outcome.

We could return an empty string instead of throwing an exception, but this still violates the LSP, as the `serialize` method is expected to return a serialized message, not an empty string. It also introduces ambiguity, as it's not clear whether an empty string is the result of a successful serialization of a message with no data or an unsuccessful serialization of `InternalMessage`.

A better approach is to separate the concerns of a `Message` and a `SerializableMessage`, where only `SerializableMessages` have a `serialize` method:

```cpp
class Message {
public:
    virtual ~Message() = default;
    // other common message behaviors
};

class SerializableMessage : public Message {
public:
    virtual std::string serialize() const = 0;
};

class StartMessage : public SerializableMessage {
    // ...
};

class StopMessage : public SerializableMessage {
    // ...
};

class InternalMessage : public Message {
    // InternalMessage doesn't have serialize method now.
};
```

In this corrected design, the base `Message` class does not include a `serialize` method, and a new `SerializableMessage` class has been introduced that includes this method. This way, only messages that can be serialized will inherit from `SerializableMessage`, and we adhere to the LSP.

Adhering to the LSP allows for more flexible and maintainable code, as it enables the use of polymorphism and allows for substituting objects of a class with objects of its subclasses without affecting the overall behavior of the program. This way, the program can take advantage of the new functionality provided by the subclass while maintaining the same behavior as the superclass.

The Interface Segregation Principle

The **Interface Segregation Principle** (**ISP**) is a principle in object-oriented programming that states that a class should only implement the interfaces it uses. In other words, it suggests that interfaces should be fine-grained and client-specific rather than having a single, large, and all-encompassing interface. The ISP is based on the idea that it is better to have many small interfaces that each define a specific set of methods rather than a single large interface that defines many methods.

One of the key benefits of the ISP is that it promotes a more modular and flexible design, as it allows for the creation of interfaces that are tailored to the specific needs of a client. This way, it reduces the number of unnecessary methods that a client needs to implement, and also it reduces the risk of a client depending on methods that it does not need.

An example of the ISP can be observed when creating our example messages from MessagePack or JSON files. Following the best practices, we would create an interface providing two methods, `from_message_pack` and `from_json`.

The current implementations need to implement both methods, but what if a particular class does not need to support both options? The smaller the interface, the better. The `MessageParser` interface will be split into two separate interfaces, each requiring the implementation of either JSON or MessagePack:

```
class JsonMessageParser {
public:
  virtual std::unique_ptr<Message>
  parse(const std::vector<uint8_t>& message_pack)
    const = 0;
};

class MessagePackMessageParser {
public:
  virtual std::unique_ptr<Message>
  parse(const std::vector<uint8_t>& message_pack)
    const = 0;
};
```

This design allows for objects derived from `JsonMessageParser` and `MessagePackMessageParser` to understand how to construct themselves from JSON and MessagePack, respectively, while preserving the independence and functionality of each function. The system remains flexible as new smaller objects can still be composed to achieve the desired functionality.

Adhering to the ISP makes the code more maintainable and less prone to errors, as it reduces the number of unnecessary methods that a client needs to implement, and it also reduces the risk of a client depending on methods that it does not need.

The Dependency inversion principle

The Dependency inversion principle is based on the idea that it is better to depend on abstractions rather than on concrete implementations, as it allows for greater flexibility and maintainability. It allows the decoupling of high-level modules from low-level modules, making them more independent and less prone to changes in the low-level modules. This way, it makes it easy to change low-level implementations without affecting high-level modules and vice versa.

The DIP can be illustrated for our messaging system if we try to use all the components via another class. Let us assume that there is a class responsible for message routing. In order to build such a class, we are going to use `MessageSender` as a communication module, `Message` based classes, and `MessageSaver`:

```cpp
class MessageRouter {
public:
  MessageRouter(ReceiverId id)
    : id_{id} {}

  void route(const Message& message) const {
    if (message.receiver_id() == id_) {
      handler_.handle(message);
    } else {
      try {
        sender_.send(message);
      } catch (const CommunicationError& e) {
        saver_.save(message);
      }
    }
  }

private:
  const ReceiverId id_;
  const MessageHandler handler_;
  const MessageSender sender_;
  const MessageSaver saver_;
};
```

The new class provides only one `route` method, which is called once a new message is available. The router handles the message to the `MessageHandler` class if the message's sender ID equals the router's. Otherwise, the router forwards the message to the corresponding receiver. In case the delivery of the message fails and the communication layer throws an exception, the router saves the message via `MessageSaver`. Those messages will be delivered some other time.

The only problem is that if any dependency needs to be changed, the router's code has to be updated accordingly. For example, if the application needs to support several types of senders (TCP and UDP), the message saver (filesystem versus remote) or message handler's logic changes. In order to make `MessageRouter` agnostic to such changes, we can rewrite it using the DIP principle:

```
class BaseMessageHandler {
public:
    virtual ~BaseMessageHandler() {}
    virtual void handle(const Message& message) const = 0;
};

class BaseMessageSender {
public:
    virtual ~BaseMessageSender() {}
    virtual void send(const Message& message) const = 0;
};

class BaseMessageSaver {
public:
    virtual ~BaseMessageSaver() {}
    virtual void save(const Message& message) const = 0;
};

class MessageRouter {
public:
    MessageRouter(ReceiverId id,
                  const BaseMessageHandler& handler,
                  const BaseMessageSender& sender,
                  const BaseMessageSaver& saver)
        : id_{id}, handler_{handler}, sender_{sender},
          saver_{saver} {}

    void route(const Message& message) const {
        if (message.receiver_id() == id_) {
            handler_.handle(message);
        } else {
            try {
                sender_.send(message);
            } catch (const CommunicationError& e) {
                saver_.save(message);
            }
        }
    }
```

```
private:
    ReceiverId id_;
    const BaseMessageHandler& handler_;
    const BaseMessageSender& sender_;
    const BaseMessageSaver& saver_;
};

int main() {
    auto id      = ReceiverId{42};
    auto handler = MessageHandler{};
    auto sender = MessageSender{
        Communication::get_instance()};
    auto saver =
        FilesystemMessageSaver{"/tmp/undelivered_messages"};

    auto router = MessageRouter{id, sender, saver};
}
```

In this revised version of the code, `MessageRouter` is now decoupled from specific implementations of the message handling, sending, and saving logic. Instead, it relies on abstractions represented by `BaseMessageHandler`, `BaseMessageSender`, and `BaseMessageSaver`. This way, any class that derives from these base classes can be used with `MessageRouter`, which makes the code more flexible and easier to extend in the future. The router is not concerned with the specifics of how messages are handled, sent, or saved – it only needs to know that these operations can be performed.

Adhering to the DIP makes code more maintainable and less prone to errors. It decouples high-level modules from low-level modules, making them more independent and less prone to changes in low-level modules. It also allows for greater flexibility, making it easy to change low-level implementations without affecting high-level modules and vice versa. Later in this book, dependency inversion will help us mock parts of the system while developing unit tests.

The KISS principle

The KISS principle, which stands for "Keep It Simple, Stupid," is a design philosophy that emphasizes the importance of keeping things simple and straightforward. This principle is particularly relevant in the world of programming, where complex code can lead to bugs, confusion, and slow development time.

Here are some examples of how the KISS principle can be applied in C++:

- **Avoid Overcomplicating Code**: In C++, it's easy to get carried away with complex algorithms, data structures, and design patterns. However, these advanced techniques can lead to code that is harder to understand and debug. Instead, try to simplify the code as much as possible. For example, using a simple `for` loop instead of a complex algorithm can often be just as effective and much easier to understand.

- **Keep Functions Small**: Functions in C++ should be small, focused, and easy to understand. Complex functions can quickly become difficult to maintain and debug, so try to keep functions as simple and concise as possible. A good rule of thumb is to aim for functions that are no longer than 30-50 lines of code.

- **Use Clear and Concise Variable Names**: In C++, variable names play a crucial role in making code readable and understandable. Avoid using abbreviations and instead opt for clear and concise names that accurately describe the purpose of the variable.

- **Avoid Deep Nesting**: Nested loops and conditional statements can make code hard to read and follow. Try to keep the nesting levels as shallow as possible, and consider breaking up complex functions into smaller, simpler functions.

- **Write Simple, Readable Code**: Above all, aim to write code that is easy to understand and follow. This means using clear and concise language and avoiding complicated expressions and structures. Code that is simple and easy to follow is much more likely to be maintainable and bug-free.

- **Avoid Complex Inheritance Hierarchy**: Complex inheritance hierarchies can make code more difficult to understand, debug, and maintain. The more complex the inheritance structure, the harder it becomes to keep track of the relationships between classes and determine how changes will affect the rest of the code.

In conclusion, the KISS principle is a simple and straightforward design philosophy that can help developers write clear, concise, and maintainable code. By keeping things simple, developers can avoid bugs and confusion and speed up development time.

The KISS and SOLID Principles together

The SOLID principles and the KISS principle are both important design philosophies in software development, but they can sometimes contradict each other.

The SOLID principles are a set of five principles that guide the design of software, aimed at making it more maintainable, scalable, and flexible. They focus on creating a clean, modular architecture that follows good object-oriented design practices.

The KISS principle, on the other hand, is all about keeping things simple. It advocates for straightforward, simple solutions, avoiding complex algorithms and structures that can make code hard to understand and maintain.

While both SOLID and KISS aim to improve software quality, they can sometimes be at odds. For example, following the SOLID principles may result in code that is more complex and harder to understand to achieve greater modularity and maintainability. Similarly, the KISS principle may result in less flexible and scalable code to keep it simple and straightforward.

In practice, developers often have to strike a balance between the SOLID principles and the KISS principle. On the one hand, they want to write code that is maintainable, scalable, and flexible. On the other hand, they want to write code that is simple and easy to understand. Finding this balance requires careful consideration of trade-offs and an understanding of when each approach is most appropriate.

When I have to choose between the SOLID and KISS approaches, I think about something my boss, Amir Taya, said, "When building a Ferrari, you need to start from a scooter." This phrase is an exaggerated example of KISS: if you do not know how to build a feature, make the simplest working version (KISS), re-iterate, and extend the solution using SOLID principles if needed.

Side effects and immutability

Side effects and immutability are two important concepts in programming that have a significant impact on the quality and maintainability of code.

Side effects refer to changes that occur in the state of the program as a result of executing a particular function or piece of code. Side effects can be explicit, such as writing data to a file or updating a variable, or implicit, such as modifying the global state or causing unexpected behavior in other parts of the code.

Immutability, on the other hand, refers to the property of a variable or data structure that cannot be modified after it has been created. In functional programming, immutability is achieved by making data structures and variables constant and avoiding side effects.

The importance of avoiding side effects and using immutable variables lies in the fact that they make code easier to understand, debug, and maintain. When code has few side effects, it is easier to reason about what it does and what it does not do. This makes finding and fixing bugs and making changes to the code easier without affecting other parts of the system.

In contrast, code with many side effects is harder to understand, as the state of the program can change in unexpected ways. This makes it more difficult to debug and maintain and can lead to bugs and unexpected behavior.

Functional programming languages have long emphasized the use of immutability and the avoidance of side effects, but it is now possible to write code with these properties using C++. The easiest way to achieve it is to follow the **C++ Core Guidelines for Constants and Immutability**.

Con.1 – by default, make objects immutable

You can declare a built-in data type or an instance of a user-defined data type as constant, resulting in the same effect. Attempting to modify it will result in a compiler error:

```
struct Data {
  int val{42};
};
```

```
int main() {
  const Data data;
  data.val = 43; // assignment of member 'Data::val' in
                 // read-only object
  const int val{42};
  val = 43; // assignment of read-only variable 'val'
}
```

The same applies to loops:

```
for (const int i : array) {
  std::cout << i << std::endl; // just reading: const
}


for (int i : array) {
  std::cout << i << std::endl; // just reading: non-const
}
```

This approach allows the prevention of hard-to-notice changes of value.

Probably, the only exception is function parameters passed by value:

```
void foo(const int value);
```

Such parameters are rarely passed as const and rarely mutated. In order to avoid confusion, it is recommended not to enforce this rule in such cases.

Con.2 – by default, make member functions const

A member function (method) shall be marked as const unless it changes the observable state of an object. The reason behind this is to give a more precise statement of design intent, better readability, maintainability, more errors caught by the compiler, and theoretically more optimization opportunities:

```
class Book {
public:
  std::string name() { return name_; }

private:
  std::string name_;
};

void print(const Book& book) {
  cout << book.name()
       << endl; // ERROR: 'this' argument to member
                // function
```

```
                        // 'name' has type 'const Book', but
                        // function is not marked
                        // const clang(member_function_call_bad_cvr)
    }
```

There are two types of constness: **physical** and **logical**:

Physical constness: An object is declared const and cannot be changed.

Logical constness: An object is declared const but can be changed.

Logical constness can be achieved with the mutable keyword. In general, it is a rare use case. The only good example I can think of is storing in an internal cache or using a mutex:

```
class DataReader {
public:
  Data read() const {
    auto lock = std::lock_guard<std::mutex>(mutex);
    // read data
    return Data{};
  }

private:
  mutable std::mutex mutex;
};
```

In this example, we need to change the mutex variable to lock it, but this does not affect the logical constness of the object.

Please be aware that there exist legacy codes/libraries that provide functions that declare T*, despite not making any changes to the T. This presents an issue for individuals who are trying to mark all logically constant methods as const. In order to enforce constness, you can do the following:

- Update the library/code to be const-correct, which is the preferred solution.
- Provide a wrapper function casting away the constness.

Example

```
void read_data(int* data); // Legacy code: read_data does
                           // not modify `*data`

void read_data(const int* data) {
  read_data(const_cast<int*>(data));
}
```

Note that this solution is a patch that can be used only when the declaration of read_data cannot be modified.

Con.3 – by default, pass pointers and references to const

This one is easy; it is far easier to reason about programs when called functions do not modify state.

Let us look at the two following functions:

```
void foo(char* p);
void bar(const char* p);
```

Does the foo function modify the data the p pointer points to? We cannot answer by looking at the declaration, so we assume it does by default. However, the bar function states explicitly that the content of p will not be changed.

Con.4 – use const to define objects with values that do not change after construction

This rule is very similar to the first one, enforcing the constness of objects that are not expected to be changed in the future. It is often helpful with classes such as Config that are created at the beginning of the application and not changed during its lifetime:

```
class Config {
public:
  std::string hostname() const;
  uint16_t port() const;
};

int main(int argc, char* argv[]) {
  const Config config = parse_args(argc, argv);
  run(config);
}
```

Con.5 – use constexpr for values that can be computed at compile time

Declaring variables as constexpr is preferred over const if the value is computed at compile time. It provides such benefits as better performance, better compile-time checking, guaranteed compile-time evaluation, and no possibility of race conditions.

Constness and data races

Data races occur when multiple threads access a shared variable simultaneously, and at least one tries to modify it. There are synchronization primitives such as mutexes, critical sections, spinlocks, and semaphores, allowing the prevention of data races. The problem with these primitives is that they either do expensive system calls or overuse the CPU, which makes the code less efficient. However, if none of the threads modifies the variable, there is no place for data races. We learned that `constexpr` is thread-safe (does not need synchronization) because it is defined at compile time. What about `const`? It can be thread-safe under the below conditions.

The variable has been `const` since its creation. If a thread has direct or indirect (via a pointer or reference) non-const access to the variable, all the readers need to use mutexes. The following code snippet illustrates constant and non-constant access from multiple threads:

```
void a() {
  auto value = int{42};
  auto t = std::thread([&]() { std::cout << value; });
  t.join();
}

void b() {
  auto value = int{42};
  auto t = std::thread([&value = std::as_const(value)]() {
    std::cout << value;
  });
  t.join();
}

void c() {
  const auto value = int{42};
  auto t = std::thread([&]() {
    auto v = const_cast<int&>(value);
    std::cout << v;
  });
  t.join();
}

void d() {
  const auto value = int{42};
  auto t = std::thread([&]() { std::cout << value; });
  t.join();
}
```

In the a function, the `value` variable is owned as non-constant by both the main thread and `t`, which makes the code potentially not thread-safe (if a developer decides to change the `value` later in the main thread). In the b, the main thread has "write" access to `value` while `t` receives it via a `const` reference, but still, it is not thread-safe. The c function is an example of very bad code: the `value` is created as a constant in the main thread and passed as a `const` reference but then the constness is cast away, which makes this function not thread-safe. Only the d function is thread-safe because neither the main thread nor `t` can modify the variable.

The data type and all sub-types of the variable are either physically constant or their logical constness implementation is thread-safe. For example, in the following example, the `Point` struct is physically constant because its `x` and `y` field members are primitive integers, and both threads have only `const` access to it:

```
struct Point {
  int x;
  int y;
};

void foo() {
  const auto point = Point{.x = 10, .y = 10};
  auto t           = std::thread([&]() { std::cout <<
    point.x; });
  t.join();
}
```

The `DataReader` class that we saw earlier is logically constant because it has a mutable variable, `mutex`, but this implementation is also thread-safe (due to the lock):

```
class DataReader {
public:
  Data read() const {
    auto lock = std::lock_guard<std::mutex>(mutex);
    // read data
    return Data{};
  }

private:
  mutable std::mutex mutex;
};
```

However, let us look into the following case. The `RequestProcessor` class processes some heavy requests and caches the results in an internal variable:

```cpp
class RequestProcessor {
public:
  Result process(uint64_t request_id,
                 Request request) const {
    if (auto it = cache_.find(request_id); it !=
      cache_.cend()) {
      return it->second;
    }
    // process request
    // create result
    auto result = Result{};
    cache_[request_id] = result;
    return result;
  }

private:
  mutable std::unordered_map<uint64_t, Result> cache_;
};

void process_request() {
  auto requests = std::vector<std::tuple<uint64_t,
    Request>>{};
  const auto processor = RequestProcessor{};
  for (const auto& request : requests) {
    auto t = std::thread([&] () {
      processor.process(std::get<0>(request),
                        std::get<1>(request));
    });
    t.detach();
  }
}
```

This class is logically safe, but the `cache_` variable is changed in a non-thread-safe way, which makes the class non-thread-safe even when declared as `const`.

Note that when working with STL containers, it is essential to remember that, despite current implementations tending to be thread-safe (physically and logically), the standard provides very specific thread-safety guarantees.

All functions in a container can be called simultaneously by various threads on different containers. Broadly, functions from the C++ standard library don't read objects accessible to other threads unless they are reachable through the function arguments, which includes the `this` pointer.

All `const` member functions are thread-safe, meaning they can be invoked simultaneously by various threads on the same container. Furthermore, the `begin()`, `end()`, `rbegin()`, `rend()`, `front()`, `back()`, `data()`, `find()`, `lower_bound()`, `upper_bound()`, `equal_range()`, `at()`, and `operator[]` (except in associative containers) member functions also behave as `const` with regard to thread safety. In other words, they can also be invoked by various threads on the same container. Broadly, C++ standard library functions won't modify objects unless those objects are reachable, directly or indirectly, via the function's non-const arguments, which includes the `this` pointer.

Different elements in the same container can be altered simultaneously by different threads, with the exception of `std::vector<bool>` elements. For example, a `std::vector` of `std::future` objects can receive values from multiple threads at once.

Operations on iterators, such as incrementing an iterator, read the underlying container but don't modify it. These operations can be performed concurrently with operations on other iterators of the same container, with the `const` member functions, or with reads from the elements. However, operations that invalidate any iterators modify the container and must not be performed concurrently with any operations on existing iterators, even those that are not invalidated.

Elements of the same container can be altered concurrently with those member functions that don't access these elements. Broadly, C++ standard library functions won't read objects indirectly accessible through their arguments (including other elements of a container) except when required by its specification.

Lastly, operations on containers (as well as algorithms or other C++ standard library functions) can be internally parallelized as long as the user-visible results remain unaffected. For example, `std::transform` can be parallelized, but `std::for_each` cannot, as it is specified to visit each element of a sequence in order.

The idea of having a single mutable reference to an object became one of the pillars of the Rust programming language. This rule is in place to prevent data races, which occur when multiple threads access the same mutable data concurrently, resulting in unpredictable behavior and potential crashes. By allowing only one mutable reference to an object at a time, Rust ensures that concurrent access to the same data is properly synchronized and avoids data races.

In addition, this rule helps prevent mutable aliasing, which occurs when multiple mutable references to the same data exist simultaneously. Mutable aliasing can lead to subtle bugs and make code difficult to reason about, especially in large and complex code bases. By allowing only one mutable reference to an object, Rust avoids mutable aliasing and helps ensure that code is correct and easy to understand.

However, it's worth noting that Rust also allows multiple immutable references to an object, which can be useful in scenarios where concurrent access is necessary but mutations are not. By allowing multiple immutable references, Rust can provide better performance and concurrency while still maintaining safety and correctness.

Summary

In this chapter, we covered the SOLID principles, the KISS principle, constness, and immutability. Let's see what you learned!

- SOLID principles: SOLID is a set of five principles that help us create code that's easy to maintain, scalable, and flexible. By understanding these principles, you're on your way to designing code that's a dream to work with!

- The KISS principle: The KISS principle is all about keeping things simple. By following this principle, you'll avoid overcomplicating your code, making it easier to maintain and debug.

- Constness: Constness is a property in C++ that makes objects read-only. By declaring objects as `const`, you can ensure that their values won't be accidentally changed, making your code more stable and predictable.

- Immutability: Immutability is all about making sure objects can't be changed after their creation. By making objects immutable, you can avoid sneaky bugs and make your code more predictable.

With these design principles under your belt, you're on your way to writing code that's both robust and reliable. Happy coding!

In the next chapter, we will try to understand what causes bad code.

3

Causes of Bad Code

In the previous chapters, we discussed coding standards in C++ and the core development principles. As we delve into refactoring existing code, it is crucial to understand what leads to subpar or bad code. Recognizing these causes enables us to avoid repeating the same mistakes, address existing issues, and prioritize future improvements effectively.

Bad code can result from various factors, ranging from external pressures to internal team dynamics. One significant factor is the need to deliver the product quickly, especially in fast-paced environments such as start-ups. Here, the pressure to release features rapidly often leads to compromises in code quality as developers might cut corners or skip essential best practices to meet tight deadlines.

Another contributing factor is the multiple ways of solving the same problem in C++. The language's flexibility and richness, while powerful, can result in inconsistencies and difficulties in maintaining a coherent code base. Different developers might approach the same problem in various ways, leading to a fragmented and harder-to-maintain code base.

The developer's personal taste also plays a role. Individual preferences and coding styles can impact the overall quality and readability of the code. What one developer considers elegant, another might find convoluted, leading to subjective differences that affect code consistency and clarity.

Lastly, a lack of knowledge of modern C++ features can result in inefficient or error-prone code. As C++ evolves, it introduces new features and paradigms that require a deep understanding to be used effectively. When developers are not up to date with these advancements, they might fall back on outdated practices, missing out on improvements that can enhance code quality and performance.

By exploring these aspects, we aim to provide a thorough understanding of the factors contributing to bad code. This knowledge is essential for any developer aiming to refactor and improve an existing code base effectively. Let's dive in and uncover the root causes of bad code in C++ development.

The need to deliver the product

When developers examine pre-existing code, they may question why it was written in a manner that is less elegant or lacks extensibility. It is often easy to criticize the job done by others, but it is crucial to understand the original developer's circumstances. Suppose the project was originally developed in a start-up company. In that case, it is important to consider that start-up culture significantly emphasizes fast product delivery and the need to outpace competitors. While this can be advantageous, it can also lead to the development of bad code. One of the main reasons for this is the pressure to deliver quickly, which may cause developers to cut corners or skip essential coding practices (for example, the SOLID principles mentioned in previous chapters) in order to meet deadlines. This can result in code that lacks proper documentation, is difficult to maintain, and may be susceptible to errors.

Furthermore, the limited resources and small development teams of start-ups can exacerbate the need for speed, as developers may not have the manpower to focus on optimizing and refining the code base. As a result, the code can become cluttered and inefficient, leading to decreased performance and increased bugs.

In addition, the focus on fast delivery in start-up culture can make it difficult for developers to keep up with the latest advancements in C++. This may result in outdated code that lacks important features, uses inefficient or deprecated functions, and is not optimized for performance.

The developer's personal taste

Another significant factor contributing to bad code is the developer's personal taste. Individual preferences and coding styles can vary widely, leading to subjective differences that impact code consistency and readability. For example, consider two developers, Bob and Alice. Bob prefers using concise, compact code that leverages advanced C++ features, while Alice favors more explicit and verbose code, prioritizing clarity and simplicity.

Bob might write a function using modern C++ features such as lambda expressions and the `auto` keyword:

```
auto process_data = [](const std::vector<int>& data) {
    return std::accumulate(data.begin(), data.end(), 0L);
};
```

Alice, on the other hand, might prefer a more traditional approach, avoiding lambdas and using explicit types:

```
long process_data(const std::vector<int>& data) {
    long sum = 0;
    for (int value : data) {
        sum += value;
    }
    return sum;
}
```

While both approaches are valid and achieve the same result, the difference in style can lead to confusion and inconsistency within the code base. If Bob and Alice are working on the same project without adhering to a common coding standard, the code can become a patchwork of differing styles, making it harder to maintain and understand.

Additionally, Bob's use of modern features might introduce complexity that could be difficult for team members unfamiliar with these features, while Alice's verbose style might be seen as overly simplistic and inefficient by those who prefer more concise code. These differences, rooted in personal taste, underscore the importance of establishing and following team-wide coding standards to ensure consistency and maintainability in the code base.

By recognizing and addressing the impact of personal coding preferences, teams can work toward creating a cohesive and readable code base that aligns with best practices and enhances overall code quality.

Multiple ways of solving the same problem in C++

C++ is a versatile language that offers multiple ways to solve the same problem, a characteristic that can both empower and confuse developers. This flexibility often leads to inconsistencies within a code base, especially when different developers have varying levels of expertise and preferences. In this chapter, we will show a few examples to illustrate how the same problem can be approached in different ways, highlighting the potential benefits and pitfalls of each method. As discussed in the *The developer's personal taste* section, developers such as Bob and Alice might approach the same problem using different techniques, leading to a fragmented code base.

Revisiting Bob and Alice's example

To recap, Bob used modern C++ features such as lambda expressions and `auto` to process data concisely, while Alice preferred a more explicit and verbose approach. Both methods achieve the same result, but the difference in style can lead to confusion and inconsistency within the code base. While Bob's approach is more compact and leverages modern C++ features, Alice's method is straightforward and easier to understand for those unfamiliar with lambdas.

Raw pointers and C functions versus Standard Library functions

Consider a project that heavily uses raw pointers and C functions for copying data, a common practice in older C++ code bases:

```
void copy_array(const char* source, char* destination, size_t size) {
    for (size_t i = 0; i < size; ++i) {
        destination[i] = source[i];
    }
}
```

This approach, while functional, is prone to errors such as buffer overflows and requires manual memory management. In contrast, a modern C++ approach would use standard library functions such as std::copy:

```
void copy_array(const std::vector<char>& source, std::vector<char>&
destination) {
    std::copy(source.begin(), source.end(), std::back_
inserter(destination));
}
```

Using std::copy not only simplifies the code but also leverages well-tested library functions that handle edge cases and improve safety.

Inheritance versus templates

Another area where C++ offers multiple solutions is code reuse and abstraction. Some projects prefer using inheritance, which can lead to a rigid and complex hierarchy:

```
class Shape {
public:
    virtual void draw() const = 0;
    virtual ~Shape() = default;
};

class Circle : public Shape {
public:
    void draw() const override {
        // Draw circle
    }
};

class Square : public Shape {
public:
    void draw() const override {
        // Draw square
    }
};

class ShapeDrawer {
public:
    explicit ShapeDrawer(std::unique_ptr<Shape> shape) : shape_
(std::move(shape)) {}

    void draw() const {
        shape_->draw();
```

```
    }

private:
    std::unique_ptr<Shape> shape_;
};
```

While inheritance provides a clear structure and allows polymorphic behavior, it can become cumbersome as the hierarchy grows. An alternative approach is to use templates to achieve polymorphism without the overhead of virtual functions. Here's how templates can be used to achieve similar functionality:

```
template<typename ShapeType>
class ShapeDrawer {
public:
    explicit ShapeDrawer(ShapeType shape) : shape_(std::move(shape))
{}

    void draw() const {
        shape_.draw();
    }

private:
    ShapeType shape_;
};

class Circle {
public:
    void draw() const {
        // Draw circle
    }
};

class Square {
public:
    void draw() const {
        // Draw square
    }
};
```

In this example, ShapeDrawer uses templates to achieve polymorphic behavior. ShapeDrawer can work with any type that provides a draw method. This approach avoids the overhead associated with virtual function calls and can be more efficient, especially in performance-critical applications.

Example – handling errors

Another example of solving the same problem in different ways is error handling. Consider a project where Bob uses a traditional error code:

```
int process_file(const std::string& filename) {
    FILE* file = fopen(filename.c_str(), "r");
    if (!file) {
        return -1; // Error opening file
    }
    // Process file
    return fclose(file);
}
```

Alice, on the other hand, prefers using exceptions for error handling:

```
void process_file(const std::string& filename) {
    std::ifstream file(filename);
    if (!file) {
        throw std::runtime_error("Error opening file");
    }
    // Process file
}
```

Using exceptions can make the code cleaner by separating error handling from the main logic, but it requires an understanding of exception safety and handling. Error code, while simpler, can clutter the code with repetitive checks and may be less informative.

Projects using different approaches

In real-world projects, you might encounter a mix of these approaches, reflecting the varied backgrounds and preferences of different developers, such as these examples:

- **Project A** uses raw pointers and C functions for performance-critical sections, relying on the developers' expertise to manage memory safely

- **Project B** prefers standard library containers and algorithms, prioritizing safety and readability over raw performance

- **Project C** employs a deep inheritance hierarchy to model its domain, emphasizing clear relationships between entities

- **Project D** utilizes templates extensively to achieve high performance and flexibility, despite the steeper learning curve and potential complexity

Each approach has its pros and cons, and choosing the right one depends on the project's requirements, the team's expertise, and the specific problem being solved. However, these multiple ways of solving the same problem can lead to a fragmented and inconsistent code base if not managed carefully.

C++ provides multiple ways to solve the same problem, from raw pointers and C functions to standard library containers and templates. While this flexibility is powerful, it can also lead to inconsistencies and complexity in a code base. Understanding the strengths and weaknesses of each approach and striving for consistency through coding standards and team agreements is crucial for maintaining high-quality, maintainable code. By embracing modern C++ features and best practices, developers can write code that is both efficient and robust, reducing the likelihood of errors and improving overall code quality.

Lack of knowledge in C++

One of the major contributors to bad code is a lack of knowledge in C++. C++ is a complex and evolving language with a wide range of features, and staying updated with its latest standards requires continuous learning. Developers who are not familiar with modern C++ practices can inadvertently write inefficient or error-prone code. This section explores how gaps in understanding C++ can lead to various issues, using examples to illustrate common pitfalls.

Consider two developers, Bob and Alice. Bob has extensive experience with older versions of C++ but hasn't kept up with recent updates, while Alice is well versed in modern C++ features.

Using raw pointers and manual memory management

Bob might use raw pointers and manual memory management, a common practice in older C++ code:

```
void process() {
    int* data = new int[100];
    // ... perform operations on data
    delete[] data;
}
```

This approach is prone to errors such as memory leaks and undefined behavior if `delete[]` is missed or incorrectly matched with new. For instance, if an exception is thrown after the allocation but before `delete[]`, the memory will leak. Alice, familiar with modern C++, would use `std::vector` to manage memory safely and efficiently:

```
void process() {
    std::vector<int> data(100);
    // ... perform operations on data
}
```

Using `std::vector` eliminates the need for manual memory management, reducing the risk of memory leaks and making the code more robust and easier to maintain.

Incorrect use of smart pointers

Bob tries to adopt modern practices but misuses `std::shared_ptr`, leading to potential performance issues:

```
std::shared_ptr<int> create() {
    std::shared_ptr<int> ptr(new int(42));
    return ptr;
}
```

This approach involves two separate allocations: one for the integer and another for the control block of `std::shared_ptr`. Alice, knowing the benefits of `std::make_shared`, uses it to optimize memory allocation:

```
std::shared_ptr<int> create() {
    return std::make_shared<int>(42);
}
```

`std::make_shared` combines the allocations into a single memory block, improving performance and cache locality.

Efficient use of move semantics

Bob might not fully understand move semantics and how they can improve performance when dealing with temporary objects. Consider a function that appends elements to `std::vector`:

```
void append_data(std::vector<int>& target, const std::vector<int>&
source) {
    for (const int& value : source) {
        target.push_back(value); // Copies each element
    }
}
```

This approach involves copying each element from `source` to `target`, which can be inefficient. Alice, understanding move semantics, would optimize this by using `std::move`:

```
void append_data(std::vector<int>& target, std::vector<int>&& source)
{
    for (int& value : source) {
        target.push_back(std::move(value)); // Moves each element
    }
}
```

By using `std::move`, Alice ensures that each element is moved rather than copied, which is more efficient. Additionally, Alice might also consider using `std::move` for the entire container if `source` is no longer needed:

```
void append_data(std::vector<int>& target, std::vector<int>&& source)
{
    target.insert(target.end(), std::make_move_iterator(source.
begin()), std::make_move_iterator(source.end()));
}
```

This approach moves the elements of the entire container efficiently, leveraging move semantics to avoid unnecessary copying.

Misusing const correctness

Bob might neglect const correctness, leading to potential bugs and unclear code:

```
class MyClass {
public:
    int get_value() { return value; }
    void set_value(int v) { value = v; }
private:
    int value;
};
```

Without const correctness, it's unclear whether `get_value` modifies the state of the object. Alice applies const correctness to clarify the intent and improve safety:

```
class MyClass {
public:
    int get_value() const { return value; }
    void set_value(int v) { value = v; }
private:
    int value;
};
```

Marking `get_value` as `const` guarantees that it does not modify the object, making the code clearer and preventing accidental modifications.

Inefficient string handling

Bob might handle strings using C-style character arrays, which can lead to buffer overflows and complex code:

```
char message[100];
strcpy(message, "Hello, world!");
std::cout << message << std::endl;
```

This approach is error-prone and difficult to manage. Alice, aware of the capabilities of std::string, simplifies the code and avoids potential errors:

```
std::string message = "Hello, world!";
std::cout << message << std::endl;
```

Using std::string provides automatic memory management and a rich set of functions for string manipulation, making the code safer and more expressive.

Undefined behavior with lambdas

Lambda functions introduced in C++11 provide powerful capabilities, but they can lead to undefined behavior if not used correctly. Bob might write a lambda that captures a local variable by reference and returns it, leading to dangling references:

```
auto create_lambda() {
    int value = 42;
    return [&]() { return value; };
}

auto lambda = create_lambda();
int result = lambda(); // Undefined behavior
```

Alice, understanding the risks, captures the variable by value to ensure it remains valid:

```
auto create_lambda() {
    int value = 42;
    return [=]() { return value; };
}

auto lambda = create_lambda();
int result = lambda(); // Safe
```

Capturing by value avoids the risk of dangling references and ensures the lambda remains safe to use.

Misunderstanding undefined behavior

Bob might inadvertently write code that leads to undefined behavior by relying on uninitialized variables:

```
int sum() {
    int x;
    int y = 5;
    return x + y; // Undefined behavior: x is uninitialized
}
```

Accessing uninitialized variables can lead to unpredictable behavior and hard-to-debug issues. Alice, understanding the importance of initialization, ensures that all variables are properly initialized:

```
int sum() {
    int x = 0;
    int y = 5;
    return x + y; // Defined behavior
}
```

Properly initializing variables prevents undefined behavior and makes the code more reliable.

Misuse of C-style arrays

Using C-style arrays can lead to various issues, such as a lack of bounds checking and difficulty in managing array sizes. Consider the following example where a function creates a C array on the stack and returns it:

```
int* create_array() {
    int arr[5] = {1, 2, 3, 4, 5};
    return arr; // Undefined behavior: returning a pointer to a local
array
}
```

Returning a pointer to a local array leads to undefined behavior because the array goes out of scope when the function returns. A safer approach is to use `std::array`, which can be returned safely from a function. It provides the `size` method and is compatible with C++ algorithms such as `std::sort`:

```
std::array<int, 5> create_array() {
    return {1, 2, 3, 4, 5};
}
```

Using `std::array` not only avoids undefined behavior but also enhances safety and interoperability with the C++ Standard Library. For example, sorting an array becomes straightforward:

```
std::array<int, 5> arr = create_array();
std::sort(arr.begin(), arr.end());
```

Insufficient pointer usage

Modern C++ provides smart pointers such as `std::unique_ptr` and `std::shared_ptr` to manage dynamic memory more safely and efficiently. It's generally better to use `std::unique_ptr` instead of raw pointers for exclusive ownership. When multiple actors need to share ownership of a resource, `std::shared_ptr` can be used. However, there are common issues related to the misuse of `std::shared_ptr`.

Building std::shared_ptr

Using the constructor of `std::shared_ptr` to create an object leads to separate allocations for the control block and the managed object:

```
std::shared_ptr<int> create() {
    std::shared_ptr<int> ptr(new int(42));
    return ptr;
}
```

A better approach is to use `std::make_shared`, which combines the allocations into a single memory block, improving performance and cache locality:

```
std::shared_ptr<int> create() {
    return std::make_shared<int>(42);
}
```

Copying std::shared_ptr by value

Copying `std::shared_ptr` by value within the same thread stack is less efficient because the reference counter is atomic. It's recommended to pass `std::shared_ptr` by reference:

```
void process_shared_ptr(std::shared_ptr<int> ptr) {
    // Inefficient: copies shared_ptr by value
}

void process_shared_ptr(const std::shared_ptr<int>& ptr) {
    // Efficient: passes shared_ptr by reference
}
```

Cyclic dependencies with std::shared_ptr

Cyclic dependencies can occur when two or more std::shared_ptr instances reference each other, preventing the reference count from reaching zero and causing memory leaks. Consider the following example:

```
struct B;

struct A {
    std::shared_ptr<B> b_ptr;
    ~A() { std::cout << "A destroyed\n"; }
};

struct B {
    std::shared_ptr<A> a_ptr;
    ~B() { std::cout << "B destroyed\n"; }
};

void create_cycle() {
    auto a = std::make_shared<A>();
    auto b = std::make_shared<B>();
    a->b_ptr = b;
    b->a_ptr = a;
}
```

In this scenario, A and B reference each other, creating a cycle that prevents their destruction. This issue can be resolved using std::weak_ptr to break the cycle:

```
struct B;

struct A {
    std::weak_ptr<B> b_ptr; // Use weak_ptr to break the cycle
    ~A() { std::cout << "A destroyed\n"; }
};

struct B {
    std::shared_ptr<A> a_ptr;
    ~B() { std::cout << "B destroyed\n"; }
};

void create_cycle() {
    auto a = std::make_shared<A>();
    auto b = std::make_shared<B>();
    a->b_ptr = b;
```

```
        b->a_ptr = a;
    }
```

Checking the std::weak_ptr status

A common mistake when using std::weak_ptr is to check its status with expired() and then lock it, which is not thread-safe:

```
std::weak_ptr<int> weak_ptr = some_shared_ptr;

void check_and_use_weak_ptr() {
    if (!weak_ptr.expired()) {
        // This is not thread-safe
        auto shared_ptr = weak_ptr.lock();
        shared_ptr->do_something();
    }
}
```

The correct approach is to lock std::weak_ptr and check that the returned std::shared_ptr is not null:

```
void check_and_use_weak_ptr_correctly() {
    // This is thread-safe
    if (auto shared_ptr = weak_ptr.lock()) {
        // Use shared_ptr
        shared_ptr->do_something();
    }
}
```

Lack of knowledge in C++ can lead to various issues, from memory management errors to inefficient and unreadable code. By staying updated with modern C++ features and best practices, developers can write code that is safer, more efficient, and easier to maintain. Continuous learning and adaptation are key to overcoming these challenges and improving overall code quality. Bob and Alice's examples highlight the importance of understanding and applying modern C++ practices to avoid common pitfalls and produce high-quality code.

Summary

In this chapter, we explored various causes of bad code in C++ and how a lack of knowledge in modern C++ practices can lead to inefficient, error-prone, or undefined behavior. By examining specific examples, we highlighted the importance of continuous learning and adaptation to keep up with the evolving features of C++.

We began by discussing the pitfalls of using raw pointers and manual memory management, showing how modern C++ practices such as `std::vector` can eliminate the need for manual memory management and reduce the risk of memory leaks. The advantages of using `std::unique_ptr` for exclusive ownership and `std::shared_ptr` for shared ownership were emphasized, while common issues such as inefficient memory allocation, unnecessary copying, and cyclic dependencies were highlighted.

In the context of `std::shared_ptr`, we demonstrated the benefits of using `std::make_shared` over the constructor to reduce memory allocations and improve performance. The efficiency gained by passing `std::shared_ptr` by reference rather than by value due to the atomic reference counter was also explained. We illustrated the problem of cyclic dependencies and how `std::weak_ptr` can be used to break cycles and prevent memory leaks. The correct way to check and use `std::weak_ptr` by locking it and checking the resulting `std::shared_ptr` to ensure thread safety was also covered.

Efficient use of move semantics was discussed to optimize performance by reducing unnecessary copying of temporary objects. Using `std::move` and `std::make_move_iterator` can significantly enhance program performance. The importance of const correctness was highlighted, showing how applying `const` to methods can clarify intent and improve code safety.

We addressed the dangers of using C-style character arrays and how `std::string` can simplify string handling, reduce errors, and provide better memory management. The misuse of C-style arrays was explored, and `std::array` was presented as a safer and more robust alternative. By using `std::array`, we can avoid undefined behavior and leverage C++ Standard Library algorithms such as `std::sort`.

Finally, the proper use of lambda functions was discussed, along with the potential pitfalls of capturing variables by reference, which can lead to dangling references. Capturing variables by value ensures that the lambda remains safe to use.

Through these examples, we learned about the critical importance of adopting modern C++ features and best practices to write safer, more efficient, and maintainable code. By staying updated with the latest standards and continuously improving our understanding of C++, we can avoid common pitfalls and produce high-quality software.

4

Identifying Ideal Candidates for Rewriting – Patterns and Anti-Patterns

Refactoring is a crucial technique in software development that involves making changes to existing code to improve its structure, readability, and maintainability without altering its behavior. It is vital for several reasons.

It helps to eliminate technical debt and enhance the overall quality of the code base. Developers can achieve this by removing redundant or duplicate code, simplifying complex code, and improving code readability, resulting in more maintainable and robust software.

Refactoring facilitates future development. By restructuring code to be more modular, developers can reuse existing code more effectively, saving time and effort in future development. This makes code more flexible and adaptable to change, making it easier to add new features, fix bugs, and optimize performance.

Well-structured and maintainable code makes it easier for multiple developers to collaborate effectively on a project. Refactoring helps to standardize code practices, reduce complexity, and improve documentation, making it easier for developers to understand and contribute to a code base.

Eventually, refactoring reduces costs associated with software development in the long term. By improving code quality and maintainability, refactoring can help to reduce the time and effort required for bug fixes, updates, and other maintenance tasks.

In this chapter, we will focus on identifying good candidates for refactoring in C++ projects. However, identifying the right code segments for refactoring can be challenging, especially in large and complex systems. Therefore, it is essential to understand the factors that make code segments ideal candidates for refactoring. In this chapter, we will explore these factors and provide guidelines for identifying good candidates for refactoring in C++. We will also discuss common refactoring techniques and tools that can be used to improve the quality of C++ code.

What kind of code is worth rewriting?

Determining whether a piece of code is worth rewriting depends on several factors, including the code's maintainability, readability, performance, scalability, and adherence to best practices. Let's look at some situations where code may be worth rewriting.

Smelly code is often an indication that code needs to be rewritten. These are signs of poor design or implementation, such as long methods, large classes, duplicated code, or poor naming conventions. Addressing these code smells can improve the overall quality of the code base and make it easier to maintain in the long run.

Code that exhibits low cohesion or high coupling might be worth rewriting. Low cohesion means that the elements within a module or class are not closely related, and the module or class has too many responsibilities. High coupling refers to a high degree of dependency between modules or classes, making the code harder to maintain and modify. Refactoring such code can lead to a more modular and easier-to-understand architecture.

In the previous chapters, we discussed the importance of SOLID principles; code that violates them can also be worth rewriting.

Another reason to rewrite code is if it relies on outdated technologies, libraries, or programming practices. Such code can become increasingly difficult to maintain over time and may not take advantage of newer, more efficient methods or tools. Updating the code to use current technologies and practices can improve its performance, security, and maintainability.

Lastly, if the code has performance or scalability issues, it may be worth rewriting. This can involve optimizing algorithms, data structures, or resource management to ensure that the code runs more efficiently and can handle larger workloads.

Smelly code and its basic characteristics

Smelly code, also known as **code smells**, refers to the symptoms in a code base that suggest underlying design or implementation issues. These symptoms are not necessarily bugs but are indicators of potential problems that can make the code harder to understand, maintain, and modify. Code smells are often the result of poor coding practices or the accumulation of technical debt over time. Although code smells might not directly affect the functionality of a program, they can significantly impact the overall code quality, leading to an increased risk of bugs and a decrease in developer productivity.

One aspect of addressing smelly code involves identifying and applying appropriate design patterns. Design patterns are reusable solutions to common problems that arise in software design. They provide a proven framework for solving specific problems, allowing developers to build on the collective wisdom and experience of other developers. By applying these patterns, it is possible to refactor smelly code into a more structured, modular, and maintainable form. Let's take a look at a few examples.

The strategy pattern allows us to define a family of algorithms, encapsulate each one in a separate class, and make them interchangeable at runtime. The strategy pattern is useful for refactoring code that has multiple branches or conditions performing similar tasks with slightly different implementations.

Let's consider an example of an application that saves data using different storage strategies, such as saving to disk or a remote storage service:

```cpp
#include <iostream>
#include <fstream>
#include <string>
#include <assert>

enum class StorageType {
    Disk,
    Remote
};

class DataSaver {
public:
    DataSaver(StorageType storage_type) : storage_type_(storage_type)
{}

    void save_data(const std::string& data) const {
        switch (storage_type_) {
            case StorageType::Disk:
                save_to_disk(data);
                break;
            case StorageType::Remote:
                save_to_remote(data);
                break;
            default:
                assert(false && "Unknown storage type.");
        }
    }

    void set_storage_type(StorageType storage_type) {
        storage_type_ = storage_type;
    }

private:
    void save_to_disk(const std::string& data) const {
        // saving to disk
    }
```

```cpp
    void save_to_remote(const std::string& data) const {
        // saving data to a remote storage service.
    }

    StorageType storage_type_;
};

int main() {
    DataSaver disk_data_saver(StorageType::Disk);
    disk_data_saver.save_data("Save this data to disk.");

    DataSaver remote_data_saver(StorageType::Remote);
    remote_data_saver.save_data("Save this data to remote storage.");

    // Switch the storage type at runtime.
    disk_data_saver.set_storage_type(StorageType::Remote);
    disk_data_saver.save_data("Save this data to remote storage after
switching storage type.");

    return 0;
}
```

In this class, the `save_data` method checks the storage type on each call and uses a `switch-case` block to decide which saving method to use. This approach works, but it has some drawbacks:

- The `DataSaver` class is responsible for handling all the different storage types, making it harder to maintain and extend.

- Adding new storage types requires modifying the `DataSaver` class and the `StorageType` enumeration, increasing the risk of introducing bugs or breaking existing functionality. For example, if for some reason the wrong enum type is provided, the code aborts.

- The code is less modular and flexible compared to the strategy pattern, where behaviors are encapsulated in separate classes.

By implementing the strategy pattern, we can address these drawbacks and create a more maintainable, flexible, and extensible design for the `DataSaver` class. First, define an interface called `SaveStrategy` that represents the saving behavior:

```cpp
class SaveStrategy {
public:
    virtual ~SaveStrategy() {}
    virtual void save_data(const std::string& data) const = 0;
};
```

Next, implement concrete `SaveStrategy` classes for each storage type:

```cpp
class DiskSaveStrategy : public SaveStrategy {
public:
    void save_data(const std::string& data) const override {
        // ...
    }
};

class RemoteSaveStrategy : public SaveStrategy {
public:
    void save_data(const std::string& data) const override {
        // ...
    }
};
```

Now, create a `DataSaver` class that uses the strategy pattern to delegate its saving behavior to the appropriate `SaveStrategy` implementation:

```cpp
class DataSaver {
public:
    DataSaver(std::unique_ptr<SaveStrategy> save_strategy)
        : save_strategy_(std::move(save_strategy)) {}

    void save_data(const std::string& data) const {
        save_strategy_->save_data(data);
    }

    void set_save_strategy(std::unique_ptr<SaveStrategy> save_
strategy) {
        save_strategy_ = std::move(save_strategy);
    }

private:
    std::unique_ptr<SaveStrategy> save_strategy_;
};
```

Finally, here's an example of how to use the `DataSaver` class with different saving strategies:

```cpp
int main() {
    DataSaver disk_data_saver(std::make_unique<DiskSaveStrategy>());
    disk_data_saver.save_data("Save this data to disk.");

    DataSaver remote_data_saver(std::make_
unique<RemoteSaveStrategy>());
```

```
remote_data_saver.save_data("Save this data to remote storage.");

    // Switch the saving strategy at runtime.
    disk_data_saver.set_save_strategy(std::make_
unique<RemoteSaveStrategy>());
    disk_data_saver.save_data("Save this data to remote storage after
switching strategy.");

    return 0;
}
```

In this example, the DataSaver class uses the strategy pattern to delegate its saving behavior to different SaveStrategy implementations, allowing it to easily switch between saving to disk and saving to remote storage. This design makes the code more modular, maintainable, and flexible, allowing new storage strategies to be added with minimal changes to the existing code. Additionally, the new version of the code does not need to terminate or throw an exception on the wrong save strategy type.

Let's assume we have file parser implementations for two formats, CSV and JSON:

```
class CsvParser {
public:
    void parse_file(const std::string& file_path) {
        std::ifstream file(file_path);
        if (!file) {
            std::cerr << "Error opening file: " << file_path <<
std::endl;
            return;
        }

        std::string line;
        while (std::getline(file, line)) {
            process_line(line);
        }

        file.close();
        post_process();
    }

private:
    void process_line(const std::string& line) {
        // Implement the CSV-specific parsing logic.
        std::cout << "Processing CSV line: " << line << std::endl;
    }

    void post_process() {
```

```
            std::cout << "CSV parsing completed." << std::endl;
        }
};

class JsonParser {
public:
    void parse_file(const std::string& file_path) {
        std::ifstream file(file_path);
        if (!file) {
            std::cerr << "Error opening file: " << file_path <<
std::endl;
            return;
        }

        std::string line;
        while (std::getline(file, line)) {
            process_line(line);
        }

        file.close();
        post_process();
    }

private:
    void process_line(const std::string& line) {
        // Implement the JSON-specific parsing logic.
        std::cout << "Processing JSON line: " << line << std::endl;
    }

    void post_process() {
        std::cout << "JSON parsing completed." << std::endl;
    }
};
```

In this example, the `CsvParser` and `JsonParser` classes have separate implementations of the `parse_file` method that contain duplicate code for opening, reading, and closing the file. The format-specific parsing logic is implemented in the `process_line` and `post_process` methods.

While this design works, it has some drawbacks: the shared parsing steps are duplicated in both classes, making it harder to maintain and update the code, and adding support for new file formats requires creating new classes with similar code structures, which can lead to even more code duplication.

By implementing the template method pattern, you can address these drawbacks and create a more maintainable, extensible, and reusable design for the file parsers. The `FileParser` base class handles the common parsing steps, while the derived classes implement the format-specific parsing logic.

As in the previous example, let's start with creating an abstract base class. `FileParser` represents the general file parsing process:

```cpp
class FileParser {
public:
    void parse_file(const std::string& file_path) {
        std::ifstream file(file_path);
        if (!file) {
            std::cerr << "Error opening file: " << file_path <<
std::endl;
            return;
        }

        std::string line;
        while (std::getline(file, line)) {
            process_line(line);
        }

        file.close();
        post_process();
    }

protected:
    virtual void process_line(const std::string& line) = 0;
    virtual void post_process() = 0;
};
```

The `FileParser` class has a `parse_file` method that handles the common steps of opening a file, reading its content line by line, and closing the file. The format-specific parsing logic is implemented by the pure virtual `process_line` and `post_process` methods, which will be overridden by the derived classes.

Now, create derived classes for different file formats:

```cpp
class CsvParser : public FileParser {
protected:
    void process_line(const std::string& line) override {
        // Implement the CSV-specific parsing logic.
        std::cout << "Processing CSV line: " << line << std::endl;
    }

    void post_process() override {
        std::cout << "CSV parsing completed." << std::endl;
    }
};
```

```
class JsonParser : public FileParser {
protected:
    void process_line(const std::string& line) override {
        // Implement the JSON-specific parsing logic.
        std::cout << "Processing JSON line: " << line << std::endl;
    }

    void post_process() override {
        std::cout << "JSON parsing completed." << std::endl;
    }
};
```

In this example, the CsvParser and JsonParser classes inherit from FileParser and implement the format-specific parsing logic in the process_line and post_process methods.

Here's an example of how to use the file parsers:

```
int main() {
    CsvParser csv_parser;
    csv_parser.parse_file("data.csv");

    JsonParser json_parser;
    json_parser.parse_file("data.json");

    return 0;
}
```

By implementing the template method pattern, the FileParser class provides a reusable template for handling the common steps of file parsing while allowing derived classes to implement format-specific parsing logic. This design makes it easy to add support for new file formats without modifying the base FileParser class, leading to a more maintainable and extensible code base. It is important to note that usually, the complicated part of implementing this design pattern is to recognize the common logic between the classes. Often the implementation requires some sort of unification of the common logic.

Another helpful pattern to look at is the observer pattern. The previous chapter mentions its technical implementation details (raw, shared, or weak pointer implementation). However, in this chapter, I would like to cover its usage from a design perspective.

The observer pattern defines a one-to-many dependency between objects, allowing multiple observers to be notified when the state of the subject changes. This pattern can be beneficial when refactoring code that involves event handling or updates to multiple dependent components.

Consider a car system where an `Engine` class holds the car's current speed and RPM (revolutions per minute). There are several elements that need to know about these values, such as `Dashboard` and `Controller`. The dashboard displays the latest update from the engine and `Controller` adjusts the car's behavior based on the speed and RPM. The straightforward way to implement this is to have the `Engine` class directly call `update` methods on each display element:

```cpp
class Dashboard {
public:
    void update(int speed, int rpm) {
        // display the current speed
    }
};

class Controller {
public:
    void update(int speed, int rpm) {
        // Adjust car's behavior based on the speed and RPM.
    }
};

class Engine {
public:
    void set_dashboard(Dashboard* dashboard) {
        dashboard_ = dashboard;
    }

    void set_controller(Controller* controller) {
        controller_ = controller;
    }

    void set_measurements(int speed, int rpm) {
        speed_ = speed;
        rpm_ = rpm;
        measurements_changed();
    }

private:
    void measurements_changed() {
        dashboard_->update(_speed, rpm_);
        controller_->update(_speed, rpm_);
    }

    int speed_;
    int rpm_;
```

```
    Dashboard* dashboard_;
    Controller* controller_;
};

int main() {
    Engine engine;
    engine.set_measurements(80, 3000);

    return 0;
}
```

This code has a couple of issues:

- The Engine class is tightly coupled with Dashboard and Controller, making it difficult to add or remove other components that might be interested in the car's speed and RPM.

- The Engine class is responsible for updating the display elements directly, which complicates the code and makes it less flexible.

We can refactor the code using the observer pattern to decouple the Engine from the display elements. The Engine class will become a subject, and Dashboard and Controller will become observers:

```
class Observer {
public:
    virtual ~Observer() {}
    virtual void update(int speed, int rpm) = 0;
};

class Dashboard : public Observer {
public:
    void update(int speed, int rpm) override {
        // display the current speed
    }
};

class Controller : public Observer {
public:
    void update(int speed, int rpm) override {
        // Adjust car's behavior based on the speed and RPM.
    }
};

class Engine {
public:
    void register_observer(Observer* observer) {
```

```
        observers_.push_back(observer);
    }

    void remove_observer(Observer* observer) {
        observers_.erase(std::remove(_observers.begin(), observers_.
end(), observer), observers_.end());
    }

    void set_measurements(int speed, int rpm) {
        speed_ = speed;
        rpm_ = rpm;
        notify_observers();
    }

private:
    void notify_observers() {
        for (auto observer : observers_) {
            observer->update(_speed, _rpm);
        }
    }

    std::vector<Observer*> observers_;
    int speed_;
    int rpm_;
};
```

The following snippet demonstrates the usage of the new class hierarchy:

```
int main() {
    Engine engine;
    Dashboard dashboard;
    Controller controller;

    // Register observers
    engine.register_observer(&dashboard);
    engine.register_observer(&controller);

    // Update measurements
    engine.set_measurements(80, 3000);

    // Remove an observer
    engine.remove_observer(&dashboard);

    // Update measurements again
```

```
    engine.set_measurements(100, 3500);

    return 0;
}
```

In this example, `Dashboard` and `Controller` are registered as observers to the `Engine` subject. When the engine's speed and RPM change, `set_measurements` is called, triggering `notify_observers`, which in turn calls the `update` method on each registered observer. This allows the `Dashboard` and `Controller` to receive the updated speed and RPM values.

Then, `Dashboard` is unregistered as an observer. When the engine's speed and RPM are updated again, only the `Controller` receives the updated values.

With this setup, adding or removing observers is as simple as calling `register_observer` or `remove_observer` on the `Engine`, and there is no need to modify the `Engine` class when adding new types of observers. The `Engine` class is now decoupled from the specific observer classes, making the system more flexible and easier to maintain.

Another great pattern is the state machine. It is not a classic pattern but probably the most powerful one. State machines, also known as **Finite State Machines** (**FSMs**), are mathematical models of computation. They're used to represent and control execution flow in both hardware and software designs. A state machine has a finite number of states, and at any given time, it's in one of these states. It transitions from one state to another in response to external inputs or predefined conditions.

In the realm of hardware, state machines are frequently used in the design of digital systems, serving as the control logic for everything from tiny microcontrollers to massive **central processing units** (**CPUs**). They govern the sequence of operations, ensuring that actions happen in the correct order and that the system responds appropriately to different inputs or conditions.

In software, state machines are equally useful, particularly in systems where the program flow is influenced by a series of states and transitions between those states. Applications range from simple button debouncing in embedded systems to complex game character behavior or communication protocol management.

State machines are ideal for situations where a system has a well-defined set of states that it cycles through, and where the transitions between states are triggered by specific events or conditions. They're particularly useful in situations where the system's behavior is not just a function of the current inputs, but also of the system's history. State machines encapsulate this history in the form of the current state, making it explicit and manageable.

Using a state machine can have numerous benefits. They can simplify complex conditional logic, making it easier to understand, debug, and maintain. They also make it easy to add new states or transitions without disturbing existing code, enhancing modularity and flexibility. Furthermore, they make the system's behavior explicit and predictable, reducing the risk of unexpected behavior.

Let's consider a real-world scenario of a distributed computing system where a job is submitted to be processed. This job goes through various states, such as `Submitted`, `Queued`, `Running`, `Completed`, and `Failed`. We will model this using the `Boost.Statechart` library. `Boost.Statechart` is a C++ library that provides a framework for building state machines. It is part of the Boost libraries collection. This library facilitates the development of hierarchical state machines, allowing you to model complex systems with intricate states and transitions. It aims to make it easier to write well-structured, modular, and maintainable code when dealing with complex state logic. `Boost.Statechart` provides both compile-time and runtime checks to help ensure the correctness of the state machine's behavior.

First, we include the necessary header files and set up some namespaces:

```
#include <boost/statechart/state_machine.hpp>
#include <boost/statechart/simple_state.hpp>
#include <boost/statechart/transition.hpp>
#include <iostream>

namespace sc = boost::statechart;
```

Next, we define our events: `JobSubmitted`, `JobQueued`, `JobRunning`, `JobCompleted`, and `JobFailed`:

```
struct EventJobSubmitted : sc::event< EventJobSubmitted > {};
struct EventJobQueued : sc::event< EventJobQueued > {};
struct EventJobRunning : sc::event< EventJobRunning > {};
struct EventJobCompleted : sc::event< EventJobCompleted > {};
struct EventJobFailed : sc::event< EventJobFailed > {};
```

Then, we define our states, each state is a class that inherits from `sc::simple_state`. We will have five states: `Submitted`, `Queued`, `Running`, `Completed`, and `Failed`:

```
struct Submitted;
struct Queued;
struct Running;
struct Completed;
struct Failed;

struct Submitted : sc::simple_state< Submitted, Job > {
    typedef sc::transition< EventJobQueued, Queued > reactions;

    Submitted() { std::cout << "Job Submitted\n"; }
};

struct Queued : sc::simple_state< Queued, Job > {
    typedef sc::transition< EventJobRunning, Running > reactions;
```

```
    Queued() { std::cout << "Job Queued\n"; }
};

struct Running : sc::simple_state< Running, Job > {
    typedef boost::mpl::list<
        sc::transition< EventJobCompleted, Completed >,
        sc::transition< EventJobFailed, Failed >
    > reactions;

    Running() { std::cout << "Job Running\n"; }
};

struct Completed : sc::simple_state< Completed, Job > {
    Completed() { std::cout << "Job Completed\n"; }
};

struct Failed : sc::simple_state< Failed, Job > {
    Failed() { std::cout << "Job Failed\n"; }
};
```

Finally, we define our state machine, Job, which starts in the Submitted state.

```
struct Job : sc::state_machine< Job, Submitted > {};
```

In a main function, we can create an instance of our Job state machine and process some events:

```
int main() {
    Job my_job;
    my_job.initiate();
    my_job.process_event(EventJobQueued());
    my_job.process_event(EventJobRunning());
    my_job.process_event(EventJobCompleted());
    return 0;
}
```

This will output the following:

```
Job Submitted
Job Queued
Job Running
Job Completed
```

This simple example shows how state machines can be used to model a process with multiple states and transitions. We used events to trigger transitions between states. Another approach is to use state reactions, where a state can decide when to transition based on the conditions or the data it has.

This can be achieved using custom reactions in `Boost.Statechart`. A custom reaction is a member function that's called when an event is processed. It can decide what to do: ignore the event, consume the event without transitioning, or transition to a new state.

Let's modify the `Job` state machine to make it decide when to transition from `Running` to `Completed` or `Failed` based on the completion status of the job.

First, we will define a new event, `EventJobUpdate`, which will carry the completion status of the job:

```cpp
struct EventJobUpdate : sc::event< EventJobUpdate > {
    EventJobUpdate(bool is_complete) : is_complete(is_complete) {}
    bool is_complete;
};
```

Then, in the `Running` state, we will define a custom reaction for this event:

```cpp
struct Running : sc::simple_state< Running, Job > {
    typedef sc::custom_reaction< EventJobUpdate > reactions;

    sc::result react(const EventJobUpdate& event) {
        if (event.is_complete) {
            return transit<Completed>();
        } else {
            return transit<Failed>();
        }
    }

    Running() { std::cout << "Job Running\n"; }
};
```

Now, the `Running` state will decide when to transition to `Completed` or `Failed` based on the `is_complete` field of the `EventJobUpdate` event.

In the `main` function, we can now process the `EventJobUpdate` event:

```cpp
int main() {
    Job my_job;
    my_job.initiate();
    my_job.process_event(EventJobQueued());
    my_job.process_event(EventJobRunning());
    my_job.process_event(EventJobUpdate(true)); // The job is
complete.
```

```
        return 0;
}
```

This will output the following:

```
Job Submitted
Job Queued
Job Running
Job Completed
```

If we process EventJobUpdate with false:

```
my_job.process_event(EventJobUpdate(false)); // The job is not
complete.
```

It will output the following:

```
Job Submitted
Job Queued
Job Running
Job Failed
```

This shows how a state can decide when to transition based on the conditions or the data it has.

Logic implemented as a state machine can be easily extended by adding new states and transition rules between them. However, at some point state machines may include too many states (let's say, more than seven). Often it is a symptom of smelly code. It means that the state machine is overloaded with too many states implementing several state machines. For example, our distributed system can be implemented as a state machine itself. The system could have its own states, such as Idle, ProcessingJobs, and SystemFailure. The ProcessingJobs state will further contain the Job state machine as a sub-state machine. The System state machine can communicate with the Job sub-state machine by processing events. When the System transitions to the ProcessingJobs state, it can process an EventJobSubmitted event to start the Job sub-state machine. When the Job transitions to the Completed or Failed state, it can process an EventJobFinished event to notify the System.

First, we define the EventJobFinished event:

```
struct EventJobFinished : sc::event< EventJobFinished > {};
```

Then, in the Completed and Failed states of the Job state machine, we process the EventJobFinished event:

```
struct Completed : sc::simple_state< Completed, Job > {
    Completed() {
        std::cout << "Job Completed\n";
```

```
                context< Job >().outermost_context().process_
event(EventJobFinished());
        }
};

struct Failed : sc::simple_state< Failed, Job > {
    Failed() {
        std::cout << "Job Failed\n";
        context< Job >().outermost_context().process_
event(EventJobFinished());
    }
};
```

In the `ProcessingJobs` state of the `System` state machine, we define a custom reaction for the `EventJobFinished` event:

```
struct ProcessingJobs : sc::state< ProcessingJobs, System, Job > {
    typedef sc::custom_reaction< EventJobFinished > reactions;

    sc::result react(const EventJobFinished&) {
        std::cout << "Job Finished\n";
        return transit<Idle>();
    }

    ProcessingJobs(my_context ctx) : my_base(ctx) {
        std::cout << "System Processing Jobs\n";
        context< System >().process_event(EventJobSubmitted());
    }
};
```

In the `main` function, we can create an instance of our `System` state machine and start it:

```
int main() {
    System my_system;
    my_system.initiate();
    return 0;
}
```

This will output the following:

```
System Idle
System Processing Jobs
Job Submitted
Job Queued
Job Running
```

```
Job Completed
Job Finished
System Idle
```

This shows how the System state machine interacts with the Job sub-state machine. The System starts the Job when it transitions to the ProcessingJobs state, and the Job notifies the System when it's finished. This allows the System to manage the life cycle of the Job and react to its state changes.

This can make your state machines more flexible and dynamic.

In general, state machines are a powerful tool for managing complex behavior robustly and understandably. Despite their utility, state machines are not always the first choice for structuring code, perhaps due to the perceived complexity or lack of familiarity. However, when dealing with a system characterized by a complex web of conditional logic, considering a state machine can be a wise move. It's a powerful tool that can bring clarity and robustness to your software design, making it an essential part of the refactoring toolkit in C++ or any other language.

Anti-patterns

In contrast to design patterns, anti-patterns are common solutions to problems that turn out to be counterproductive or harmful in the long run. Recognizing and avoiding anti-patterns is crucial in addressing smelly code, as applying them can exacerbate existing issues and introduce new ones. Some examples of anti-patterns include Singleton, God Object, Copy-Paste Programming, Premature Optimization, and Spaghetti Code.

Singleton is known to violate dependency inversion and open/closed principles. It creates a global instance, which can lead to hidden dependencies between classes and make the code hard to understand and maintain. It violates the dependency inversion principle, as it encourages high-level modules to depend on low-level modules instead of depending on abstractions. Additionally, the singleton pattern often makes it difficult to replace the singleton instance with a different implementation, for example, when extending the class or during testing. This violates the open/closed principle, as it requires modifying the code to change or extend the behavior. In the following code sample, we have a singleton class, Database, used by the OrderManager class:

```cpp
class Database {
public:
    static Database& get_instance() {
        static Database instance;
        return instance;
    }

    template<typename T>
    std::optional<T> get(const Id& id) const;
```

```
    template<typename T>
    void save(const T& data);

private:
    Database() {} // Private constructor
    Database(const Database&) = delete; // Delete copy constructor
    Database& operator=(const Database&) = delete; // Delete copy
assignment operator
};

class OrderManager {
public:
  void addOrder(const Order& order) {
    auto db = Database::get_instance();
    // check order validity
    // notify other components about the new order, etc
    db.save(order);
  }
};
```

The idea of having the database connection represented as a singleton is quite logical: the application allows having a single database connection per application instance, and the database is used everywhere in the code. The usage of singleton hides the fact that OrderManager depends on Database, which makes the code less obvious and predictable. The usage of singleton makes it almost impossible to test the business logic of OrderManager via unit tests without running a real instance of the database aside.

The problem can be solved by creating an instance of Database somewhere at the beginning of the main function and passing it to all the classes that need a database connection:

```
class OrderManager {
  public:
  OrderManager(Database& db);
  // the rest of the code is the same
};

int main() {
  auto db = Database{};
  auto order_manager = OrderManager{db};
}
```

Note that despite the fact that Database is not a singleton anymore (that is, its constructor is public), it still cannot be copied. Technically, this allows developers to create new instances ad hoc, which is not a desired behavior. In my experience, it can be easily avoided by knowledge sharing within the team and enforced by code review. Those developers who think it is not enough can keep Database unchanged but make sure that get_instance is called only once and passed by reference since then:

```
int main() {
  auto db = Database::get_instance();
  auto order_manager = OrderManager{db};
}
```

If a code smell involves a class with too many responsibilities, applying the god object anti-pattern would be inappropriate, as it would only make the class more convoluted and difficult to maintain. In general, god class is a violation of the single responsibility principle on steroids. For example, let's take a look at the following class, EcommerceSystem:

```
class ECommerceSystem {
public:
    // Product management
    void add_product(int id, const std::string& name, uint64_t price)
{
        products_[id] = {name, price};
    }

    void remove_product(int id) {
        products_.erase(id);
    }

    void update_product(int id, const std::string& name, uint64_t
price) {
        products_[id] = {name, price};
    }

    void list_products() {
        // print the list of products
    }

    // Cart management
    void add_to_cart(int product_id, int quantity) {
        cart_[product_id] += quantity;
    }

    void remove_from_cart(int product_id) {
        cart_.erase(product_id);
```

```cpp
    }

    void update_cart(int product_id, int quantity) {
        cart_[product_id] = quantity;
    }

    uint64_t calculate_cart_total() {
        uint64_t total = 0;
        for (const auto& item : cart_) {
            total += products_[item.first].second * item.second;
        }
        return total;
    }

    // Order management
    void place_order() {
        // Process payment, update inventory, send confirmation email,
etc.
        // ...
        cart_.clear();
    }

    // Persistence
    void save_to_file(const std::string& file_name) {
        // serializing the state to a file
    }

    void load_from_file(const std::string& file_name) {
        // loading a file and parsing it
    }

private:
    std::map<int, std::pair<std::string, uint64_t>> products_;
    std::map<int, int> cart_;
};
```

In this example, the ECommerceSystem class takes on multiple responsibilities such as product management, cart management, order management, and persistence (saving and loading data from a file). This class is difficult to maintain, understand, and modify.

A better approach would be to break down the `ECommerceSystem` into smaller, more focused classes, each handling a specific responsibility:

- The `ProductManager` class manages products
- The `CartManager` class manages the cart
- The `OrderManager` class manages orders and related tasks (e.g., processing payments and sending confirmation emails)
- The `PersistenceManager` class handles saving and loading data from files

These classes can be implemented as follows:

```cpp
class ProductManager {
public:
    void add_product(int id, const std::string& name, uint64_t price)
{
        products_[id] = {name, price};
    }

    void remove_product(int id) {
        products_.erase(id);
    }

    void update_product(int id, const std::string& name, uint64_t
price) {
        products_[id] = {name, price};
    }

    std::pair<std::string, uint64_t> get_product(int id) {
        return products_[id];
    }

    void list_products() {
        // print the list of products
    }

private:
    std::map<int, std::pair<std::string, uint64_t>> products_;
};

class CartManager {
public:
    void add_to_cart(int product_id, int quantity) {
        cart_[product_id] += quantity;
```

```cpp
    }

    void remove_from_cart(int product_id) {
        cart_.erase(product_id);
    }

    void update_cart(int product_id, int quantity) {
        cart_[product_id] = quantity;
    }

    std::map<int, int> get_cart_contents() {
        return cart_;
    }

    void clear_cart() {
        cart_.clear();
    }

private:
    std::map<int, int> cart_;
};

class OrderManager {
public:
    OrderManager(ProductManager& product_manager, CartManager& cart_
manager)
        : product_manager_(product_manager), cart_manager_(cart_
manager) {}

    uint64_t calculate_cart_total() {
        // calculate cart's total the same as before
    }

    void place_order() {
        // Process payment, update inventory, send confirmation email,
etc.
        // ...
        cart_manager_.clear_cart();
    }

private:
    ProductManager& product_manager_;
    CartManager& cart_manager_;
};
```

```
class PersistenceManager {
public:
    PersistenceManager(ProductManager& product_manager)
        : product_manager_(product_manager) {}

    void save_to_file(const std::string& file_name) {
      // saving
    }

    void load_from_file(const std::string& file_name) {
      // loading
    }

private:
    ProductManager& product_manager_;
};
```

Eventually, the `ECommerce` class that owns the new classes and provides proxy methods to their functionality:

```
// include the new classes

class ECommerce {
public:
    void add_product(int id, const std::string& name, uint64_t price)
{
        product_manager_.add_product(id, name, price);
    }

    void remove_product(int id) {
        product_manager_.remove_product(id);
    }

    void update_product(int id, const std::string& name, uint64_t
price) {
        product_manager_.update_product(id, name, price);
    }

    void list_products() {
        product_manager_.list_products();
    }

    void add_to_cart(int product_id, int quantity) {
```

```cpp
        cart_manager_.add_to_cart(product_id, quantity);
    }

    void remove_from_cart(int product_id) {
        cart_manager_.remove_from_cart(product_id);
    }

    void update_cart(int product_id, int quantity) {
        cart_manager_.update_cart(product_id, quantity);
    }

    uint64_t calculate_cart_total() {
        return order_manager_.calculate_cart_total();
    }

    void place_order() {
        order_manager_.place_order();
    }

    void save_to_file(const std::string& filename) {
        persistence_manager_.save_to_file(filename);
    }

    void load_from_file(const std::string& filename) {
        persistence_manager_.load_from_file(filename);
    }

private:
    ProductManager product_manager_;
    CartManager cart_manager_;
    OrderManager order_manager_{product_manager_, cart_manager_};
    PersistenceManager persistence_manager_{product_manager_};
};

int main() {
    ECommerce e_commerce;

    e_commerce.add_product(1, "Laptop", 999.99);
    e_commerce.add_product(2, "Smartphone", 699.99);
    e_commerce.add_product(3, "Headphones", 99.99);

    e_commerce.list_products();

    e_commerce.add_to_cart(1, 1); // Add 1 Laptop to the cart
```

```
    e_commerce.add_to_cart(3, 2); // Add 2 Headphones to the cart

    uint64_t cart_total = e_commerce.calculate_cart_total();
    std::cout << "Cart Total: $" << cart_total << std::endl;

    e_commerce.place_order();
    std::cout << "Order placed successfully!" << std::endl;

    e_commerce.save_to_file("products.txt");
    e_commerce.remove_product(1);
    e_commerce.remove_product(2);
    e_commerce.remove_product(3);

    std::cout << "Loading products from file..." << std::endl;
    e_commerce.load_from_file("products.txt");
    e_commerce.list_products();

    return 0;
}
```

By dividing the responsibilities among multiple smaller classes, the code becomes more modular, easier to maintain, and better suited to real-life applications. Small changes in the internal business logic of one of the subclasses will not necessitate updates to the ECommerce class. In C++ it might be even more important due to the notorious compilation time issues. It is easier to test these classes separately, or completely replace the implementation of one of them, for example, to save the data not to the disk but to remote storage.

The pitfalls of magic numbers – a case study on data chunking

Let's consider the following C++ function, send, which aims to send a block of data in chunks to some destination. Here's how the function looks:

```
#include <cstddef>
#include <algorithm>

// Actually sending the data
void do_send(const std::uint8_t* data, size_t size);

void send(const std::uint8_t* data, size_t size) {
    for (std::size_t position = 0; position < size;) {
        std::size_t length = std::min(size_t{256}, size - position);
// 256 is a magic number
        do_send(data + position, position + length);
        position += length;
```

```
        }
    }
```

What does the code do?

The send function takes a pointer to a std::uint8_t array (data) and its size (size). It then proceeds to send this data in chunks to the do_send function, which is responsible for the actual sending process. The chunks have a maximum size of 256 bytes each, as defined within the send function.

Why is the magic number problematic?

The number 256 is directly embedded into the code, and there's no explanation for what it represents. This is a classic example of a **magic number**. Anyone reading this code would have to guess why 256 was chosen. Is it a hardware limit? A protocol constraint? A performance tuning parameter?

The constexpr solution

One way to improve the clarity of this code is to replace the magic number with a named constexpr variable. For instance, the code could be rewritten like this:

```
#include <cstddef>
#include <algorithm>

constexpr std::size_t MAX_DATA_TO_SEND = 256;   // Named constant
replaces magic number

// Actually sending the data
void do_send(const std::uint8_t* data, size_t size);

void send(const std::uint8_t* data, size_t size) {
    for (std::size_t position = 0; position < size;) {
        std::size_t length = std::min(MAX_DATA_TO_SEND, size -
position);   // Use the named constant
        do_send(data + position, position + length);
        position += length;
    }
}
```

Advantages of using constexpr

Replacing the magic number with MAX_DATA_TO_SEND makes it easier to understand why this limit exists. Furthermore, if you have another function, such as read, which also needs to read data in chunks of 256 bytes, using the constexpr variable ensures consistency. If ever the chunk size needs to be changed, you only have to update it in one place, thereby reducing the risk of bugs and inconsistencies.

When dealing with smelly code, it is essential to understand the underlying causes of the smells and to apply the correct patterns or avoid anti-patterns to refactor the code effectively. For instance, if a code smell involves duplicated code, one should avoid Copy-Paste Programming and instead apply patterns like the Template Method or Strategy pattern to promote code reuse and reduce duplication. Similarly, if a code smell involves tightly coupled modules or classes, you should apply patterns such as Adapter or the Dependency Inversion Principle to reduce coupling and improve modularity.

It is important to remember that refactoring smelly code should be an iterative and incremental process. Developers should continuously review and evaluate their code base for smells, making small, focused changes that gradually improve the code's quality and maintainability. This approach allows better risk management, as it minimizes the chances of introducing new bugs or issues during the refactoring process. The best way to achieve that is unit tests. They help verify that the refactored code still meets its original requirements and behaves as expected, even after modifications to its internal structure or organization. By having a strong set of tests in place before starting the refactoring process, developers can have confidence that their changes will not negatively impact the application's behavior. This allows them to focus on improving the code's design, readability, and maintainability without worrying about unintentionally breaking the functionality. We will explore unit tests in *Chapter 13*.

In conclusion, smelly code is a term that describes symptoms in a codebase that indicate potential design or implementation issues. Addressing smelly code involves recognizing and applying appropriate design patterns, as well as avoiding anti-patterns that can be detrimental to code quality. By understanding the underlying causes of code smells and using patterns and anti-patterns effectively, developers can refactor their code base to be more maintainable, readable, and resilient to future changes. Continuous evaluation and incremental refactoring are key to keeping code smells at bay and ensuring a high-quality, efficient codebase that can adapt to evolving requirements and demands.

Legacy code

Refactoring legacy C++ code is a significant undertaking that has the potential to breathe new life into an aging code base. Often, legacy code is written in old dialects of C++, such as C++98 or C++03, which do not take advantage of the new language features and standard library improvements introduced in C++11, C++14, C++17, and C++20.

One common area for modernization is memory management. Legacy C++ code often uses raw pointers for managing dynamic memory, leading to potential issues with memory leaks and null pointer dereferencing. Such code can be refactored to use smart pointers, such as `std::unique_ptr` and `std::shared_ptr`, which automatically manage the lifetime of the objects they point to, reducing the risk of memory leaks.

Another modernization opportunity lies in adopting the range-based `for` loops introduced in C++11. Older loops with explicit iterators or index variables can be replaced with cleaner and more intuitive range-based loops. This not only makes the code easier to read but also reduces the potential for off-by-one and iterator invalidation errors.

As mentioned in previous chapters, legacy C++ code bases often make heavy use of raw arrays and C-style strings. Such code can be refactored to use `std::array`, `std::vector`, and `std::string`, which are safer, more flexible, and provide useful member functions.

Lastly, modern C++ has made significant strides in improving concurrency support with the introduction of `std::thread`, `std::async`, and `std::future` in C++11, followed by further enhancements in subsequent standards. Legacy code that uses platform-specific threading or older concurrency libraries could benefit from refactoring to use these modern, portable concurrency tools.

Let's start with an example of legacy code that uses `pthread` to create a new thread. This thread will perform a simple calculation:

```cpp
#include <pthread.h>
#include <iostream>

void* calculate(void* arg) {
    int* result = new int(0);
    for (int i = 0; i < 10000; ++i)
        *result += i;
    pthread_exit(result);
}

int main() {
    pthread_t thread;
    if (pthread_create(&thread, nullptr, calculate, nullptr)) {
        std::cerr << "Error creating thread\n";
        return 1;
    }

    int* result = nullptr;
    if (pthread_join(thread, (void**)&result)) {
        std::cerr << "Error joining thread\n";
        return 2;
    }

    std::cout << "Result: " << *result << '\n';
    delete result;

    return 0;
}
```

Now, we can refactor this code using `std::async` from C++11:

```cpp
#include <future>
#include <iostream>
```

```
int calculate() {
    int result = 0;
    for (int i = 0; i < 10000; ++i)
        result += i;
    return result;
}

int main() {
    std::future<int> future = std::async(std::launch::async,
calculate);
    try {
        int result = future.get();
        std::cout << "Result: " << result << '\n';
    } catch (const std::exception& e) {
        std::cerr << "Error: " << e.what() << '\n';
        return 1;
    }
    return 0;
}
```

In the refactored version, we use `std::async` to start a new task and `std::future::get` to obtain the result. The calculate function directly returns the result as an `int`, which is much simpler and safer than allocating memory in the `pthread` version. There are a few things to note. The call to `std::future::get` blocks the execution until the async is done. Additionally, the example uses `std::launch::async`, which ensures that the task is launched in a separate thread. The C++ 11 standard allows the implementations to decide what is the default policy: a separate thread or a deferred execution. At the time of writing, Microsoft Visual C++, GCC, and Clang run the task in a separate thread by default. The only difference is that while GCC and Clang create a new thread per task, Microsoft Visual C++ reuses threads from an internal thread pool. The error handling is also simpler, as any exception thrown by the calculate function will be caught by `std::future::get`.

Often legacy code uses object-oriented wrappers around `pthread` and other platform-specific APIs. Replacing them with the standard C++ implementation can decrease the amount of code that the developers have to support and make code more portable. However, multi-threading is a complex topic, so if the existing code has some rich thread-related logic, it is important to make sure that it stays intact.

The built-in algorithms provided with modern C++ can improve legacy code readability and maintenance. Often, developers need to check whether an array contains a certain value. Pre-C++11 language allowed doing something like this:

```
#include <vector>
#include <iostream>

int main() {
```

```cpp
    std::vector<int> numbers = {1, 2, 3, 4, 5, 6};

    bool has_even = false;
    for (size_t i = 0; i < numbers.size(); ++i) {
        if (numbers[i] % 2 == 0) {
            has_even = true;
            break;
        }
    }

    if (has_even)
        std::cout << "The vector contains an even number.\n";
    else
        std::cout << "The vector does not contain any even
numbers.\n";

    return 0;
}
```

With C++11, we can use `std::any_of`, a new algorithm that checks if any element in a range satisfies a predicate. This allows us to write the code more concisely and expressively:

```cpp
#include <vector>
#include <algorithm>
#include <iostream>

int main() {
    std::vector<int> numbers = {1, 2, 3, 4, 5, 6};

    bool has_even = std::any_of(numbers.begin(), numbers.end(),
                        [](int n) { return n % 2 == 0; });

    if (has_even)
        std::cout << "The vector contains an even number.\n";
    else
        std::cout << "The vector does not contain any even
numbers.\n";

    return 0;
}
```

In this refactored version, we use a lambda function as the predicate for `std::any_of`. This makes the code more concise and the intention clearer. Algorithms such as `std::all_of``  and `std::none_of`` allows to clearly express similar checks

Remember, refactoring should be done incrementally, with each change tested thoroughly to ensure it doesn't introduce new bugs or regressions. It can be a time-consuming process, but the benefits in terms of improved code quality, maintainability, and performance can be substantial.

Summary

In this chapter, we've explored some of the key design patterns that can be instrumental in refactoring legacy C++ code, including the strategy pattern, template method pattern, and observer pattern. These patterns, when applied judiciously, can significantly improve the structure of your code, making it more flexible, maintainable, and resilient to change.

While we've provided practical, real-world examples to illustrate the use of these patterns, this is by no means an exhaustive treatment. Design patterns are a vast and deep subject, with many more patterns and variations to explore. For a more comprehensive understanding of design patterns, I strongly recommend you delve into the seminal work *Design Patterns: Elements of Reusable Object-Oriented Software* by Erich Gamma, Richard Helm, Ralph Johnson, and John Vlissides, often referred to as the *Gang of Four* book.

In addition, to keep abreast of the most recent developments and emerging best practices, consider resources such as *Hands-On Design Patterns with C++: Solve common C++ problems with modern design patterns and build robust applications* by Fedor G. Pikus, and *C++ Concurrency in Action* by Anthony Williams. These works will provide you with a broader perspective and deeper understanding of the powerful role design patterns play in crafting high-quality C++ software.

Remember, the goal of refactoring and applying design patterns is not just to write code that works, but to write code that is clean, easy to understand, easy to modify, and easy to maintain in the long run.

In the upcoming chapter, we'll be delving deeper into the world of C++, focusing specifically on naming conventions, their importance in writing clean and maintainable code, and best practices established by the community.

5
The Significance of Naming

As you delve deeper into the world of C++, or any other programming language for that matter, one thing becomes increasingly clear – the power of a name. In this chapter, we will explore the profound importance of naming conventions in writing clean, maintainable, and efficient C++ code.

In computer programming, names are given to variables, functions, classes, and numerous other entities. These names serve as identifiers, playing a pivotal role in how we, as programmers, interact with the components of our code. While it may seem a trivial matter to some, choosing the right names can have a profound impact on the understandability and maintainability of a software project. The names we choose to represent the different elements of our program are the first layer of documentation that anyone, including our future selves, has when they approach our code.

Imagine a developer named Mia who works with a class named `WeatherData`. This class has two getter methods – `get_tempreture()` and `get_humidity()`. The former method simply returns the current temperature value stored in a member variable. It's an O(1) operation, since it involves returning an already-stored value. The latter does more than just return a value. It actually initiates a connection to a remote weather service, retrieves the latest humidity data, and then returns it. This operation is considerably costly, involving network communication and data processing, making it far from an O(1) operation. Mia, focusing on optimizing a function in the project, sees these two getters and assumes they are similar in terms of efficiency due to their naming. She uses `get_humidity()` within a loop, expecting it to be a simple, efficient retrieval of a stored value, similar to `get_temperature()`. The performance of the function plummets due to the repeated, expensive calls to `get_humidity()`. The network requests and data processing involved in each call significantly slow down the execution, leading to an inefficient use of resources and a slowdown in the application's performance. This could have been avoided if the method had been named `fetch_humidity()` instead of `get_humidity()`. The name `fetch_humidity()` would have made it clear that the method is not a simple getter but, rather, a more expensive operation that involves fetching data from a remote service.

The art of naming requires careful consideration and a good understanding of both the problem domain and the programming language. This chapter provides a comprehensive discussion of the general approach to creating and naming variables, class members, methods, and functions in C++. We will debate the trade-offs of long names versus short names, as well as the role of comments in clarifying our intentions.

We will explore the importance of coding conventions and the benefits they bring to individual developers and teams alike. Consistent application of a well-thought-out naming convention can streamline the coding process, reduce errors, and greatly enhance the readability of the code base.

By the end of this chapter, you will understand why good naming practices are not just an afterthought but also an essential component of good software development. We will equip you with strategies and conventions to help you write code that can be easily read, understood, and maintained by others – and by you, when you revisit your own code months or years down the line.

General naming principles

Regardless of the specific **Object-Oriented Programming** (**OOP**) language you're using, certain universal naming principles can help improve the clarity and maintainability of your code. These principles aim to ensure that names in your code provide sufficient information about their use and functionality.

Descriptiveness

Names should accurately describe the purpose or value of the variable, function, class, or method they are identifying. For instance, `getSalary()` for a function is more informative than simply `getS()`.

Consistency

Consistency in naming conventions is one of the most vital principles in writing clear and maintainable code. When you're consistent with your naming throughout your code base, it becomes much easier to read, understand, and debug your code. The reason for this is that once a developer learns your naming pattern, they can apply their understanding across the entire code base, rather than having to figure out what each individual name means in isolation.

Consistency applies to many areas, including the following:

- **Case style**: If you start by naming your variables in `snake_case` (e.g., `employee_salary`), stick to that style throughout your entire code base. Don't switch between snake_case, camelCase (e.g., `employeeSalary`), and PascalCase (e.g., `EmployeeSalary`).
- **Prefixes and suffixes**: If you use prefixes or suffixes to denote certain properties, such as `m_` for member variables (e.g., `m_value`), ensure this rule is followed everywhere.

- **Abbreviation rules**: If you decide to abbreviate certain words, ensure you do it consistently. For example, if you're using num to represent number (as in numEmployees), then always use num whenever you mean number.

- **Naming conventions specific to language constructs**: In OOP languages, class names are typically nouns (Employee), method names are verbs (calculateSalary), and Boolean variables or methods often start with is, has, can, or similar prefixes (isAvailable and hasCompleted). Stick with these conventions consistently.

Let's say you're working on a large code base where classes represent various types of employees in a company. You've decided to name your classes as singular nouns in PascalCase, methods as verbs in snake_case, and variables in snake_case too.

A consistent implementation of this naming convention might look like this:

```
class SoftwareEngineer {
public:
    void assign_task(std::string task_name) {
        current_task_ = std::move(task_name);
    }
private:
    std::string current_task_;
};
```

Let's break down this code snippet:

- The SoftwareEngineer class is a singular noun and uses PascalCase

- The assign_task method is a verb and uses snake_case

- The variable current_task is in snake_case

Being consistent with this convention will help anyone reading your code to immediately recognize what each name represents. This way, the cognitive load is reduced, and developers can focus on the actual logic, rather than being distracted by inconsistent or confusing names.

Unambiguity

Unambiguity means that names should not be misleading. Avoid using names that could be interpreted in multiple ways, or that contradict established conventions or expectations. For example, suppose you have a Document class and a method named process. Without more context, the method name is ambiguous:

```
class Document {
public:
    void process();
};
```

In this case, `process` could mean a number of things. Are we going to parse the document? Are we going to render it? Are we going to save it to a file? Or are we going to perform all these operations? It's unclear.

A more specific method name could help to clarify its purpose. Depending on what the method is supposed to do, it could be named `parse`, `render`, `save`, and so on:

```cpp
class Document {
public:
    void parse(const std::string& content);
    void render();
    void save(const std::string& file_path);
};
```

Each of these method names gives a much clearer indication of what the method does, removing the ambiguity of the original `process` method name.

Pronounceability

Names should be easily pronounceable. This facilitates verbal communication about the code between developers.

Scope and lifetimes

Variables with larger scopes and longer lifetimes usually have more influence on a system and, hence, deserve more thoughtful, clear, and descriptive names. This helps ensure that they're understood in all contexts where they're used. Here's a more detailed breakdown.

Global variables can be accessed from anywhere in a program, and their lifetime extends for the duration of the program. Therefore, they deserve especially careful consideration when naming. The name should be descriptive enough to clearly indicate its role in the system. Additionally, global variables can create unexpected dependencies, which makes the program harder to understand and maintain. Therefore, the use of global variables should be minimized:

```cpp
// Global variable
constexpr double GRAVITATIONAL_ACCELERATION = 9.8; // Clear and
descriptive
```

Class member variables can be accessed from any method within the class, and their lifetime is tied to the lifetime of the class instance. They should have clear and descriptive names that reflect their role within the class. It's often useful to follow a naming convention that distinguishes them from local variables (e.g., an `m_` prefix or `_` suffix):

```cpp
class PhysicsObject {
    double mass_;  // Descriptive and follows naming convention
```

```
    // ...
};
```

Local variables are confined to a specific function or block and exist only for the duration of that function or block. These variables typically need less descriptive names compared to global variables or class members, but they should still clearly convey their purpose:

```
double compute_force(double mass, double acceleration) {
    double force = mass * acceleration;  // 'force' is clear in this
context
    return force;
}
```

Loop variables and temporary variables have the shortest scope and lifetime, usually confined to a small loop or a small block of code. As a result, they typically have the simplest names (such as i, j, and temp):

```
for (int i = 0; i < num; ++i) {  // 'i' is clear in this context
    // ...
}
```

The key idea here is, the broader the scope and the longer the lifetime of a variable, the more potential there is for confusion about its purpose, so the more descriptive its name should be. The goal is to make code as clear and understandable as possible.

Avoid encoding type or scope information

In modern programming languages, encoding type or scope information into names (often known as Hungarian notation) is usually unnecessary and can lead to confusion or errors, especially when refactoring. While this can occasionally be helpful, especially in languages with weak typing, it has several downsides that make it less suitable for use in strongly typed languages such as C++:

* The type of a variable might change in the future, but its name often doesn't. This leads to misleading situations where a variable's name suggests one type but it actually has another. For example, you might start with a vector of IDs (std::vector<Id> id_array) and later change it to set<Id> to avoid duplication, but the variable name still suggests it's an array or vector.

* Modern development environments provide features such as type inference, hovering tooltips showing types, and powerful refactoring tools, which all make manually encoding types into names largely redundant. For example, VS Code with the clangd plugin installed and the "inlay hints" feature turned on deducts types on the fly, including auto:

```
#include <iostream>

using Id = uint64_t;

class Task {
public:
    Task(const std::string &name, Id id) : name_{name}, id_{id} {}
    auto &name()-> const std::string & const { return name_; }
    auto id()-> Id const { return id_; }

private:
    std::string name_;
    Id id_;
};

int main() {
    auto task: Task  = Task(name: "bg_task", id: Id{0});
    auto &name: const std::string & = task.name();
    std::cout << "task name: " << name << std::endl;
    return 0;
}
```

Figure 5.1 – Inlay hints in VS Code

This applies to CLion by JetBrains too:

- Prefixes in Hungarian notation can make variable names harder to read, especially for those not familiar with the notation. It may not be immediately obvious to a new developer what dwCount (a DWORD, or double word, often used to represent an unsigned long integer) means.

- Strongly typed languages such as C++ already check type safety at compile time, reducing the need to encode type information in the variable name. In the following example, integers is declared as std::vector<int>, and sentence is declared as std::string. The C++ compiler is aware of these types and will ensure that operations on these variables are type-safe:

```
#include <vector>
#include <string>

int main() {
    std::vector<int> integers;
    std::string sentence;

    // The following will cause a compile-time error because
    // the type of 'sentence' is string, not vector<int>.
    integers = sentence;

    return 0;
}
```

When the code attempts to assign `sentence` to `integers`, a compile-time error is produced because `sentence` is not of the correct type (`std::vector<int>`). This happens despite the fact that neither of the variable names encode type information.

The compiler's type-checking eliminates the need to include the type in the variable names (such as `strSentence` or `vecIntegers`), a practice common in languages that do not perform such strong compile-time type-checking. The `integers` and `sentence` variable names are sufficiently descriptive without encoding the type information.

In programming, you often come across situations where multiple logical concepts are represented using the same underlying type. For instance, you may have identifiers for `Users` and `Products` in your system, both of which are represented as integers. While C++'s static type checking provides a level of safety, it won't differentiate between `UserId` and `ProductId` – they're both just integers as far as the compiler is concerned.

However, using the same type for these different concepts can lead to bugs. It's entirely possible, for example, to mistakenly pass `UserId` where `ProductId` was expected, and the compiler wouldn't catch this error.

To address this issue, you can leverage C++'s rich type system to introduce new types representing these different concepts, even when they share the same underlying representation. This way, the compiler can catch these bugs at compile time, enhancing the robustness of your software:

```cpp
// Define new types for User and Product IDs.
struct UserId {
    explicit UserId(int id): value(id) {}
    int value;
};

struct ProductId {
    explicit ProductId(int id): value(id) {}
    int value;
};

void process_user(UserId id) {
    // Processing user...
}

void process_product(ProductId id) {
    // Processing product...
}

int main() {
    UserId user_id(1);
    ProductId product_id(2);
```

```
    // The following line would cause a compile-time error because
    // a ProductId is being passed to process_user.
    process_user(product_id);

    return 0;
}
```

In the preceding example, `UserId` and `ProductId` are distinct types. Even though their underlying representation is the same (`int`), passing `ProductId` to a function expecting `UserId` results in a compile-time error. This adds an additional layer of type safety to your code.

This is just a glimpse into how you can utilize C++'s rich static type system to create more robust and safer code. We will delve into this topic in more detail in *Chapter 6, Utilizing a Rich Static Type System in C++*.

Class and method naming

In OOP languages, classes represent concepts or things, and their instances (objects) are specific manifestations of those things. As such, both class names and their instances are most appropriately named using nouns or noun phrases. They represent entities in the system, whether they're tangible (such as `Employee` and `Invoice`) or conceptual (such as `Transaction` and `DatabaseConnection`).

On the other hand, methods in classes typically represent actions that an object of that class can perform, or messages that can be sent to it. As such, they're most effectively named using verbs or verb phrases. They act as instructions that can be carried out by the object, allowing it to interact with other objects in meaningful ways.

Consider a `Document` class with a `print` method. We can say "document, print" or "print the document," which is a clear, imperative statement in line with how we might communicate the action in everyday language.

Here's an example:

```
class Document {
public:
    void print();
};

Document report;
report.print();  // "report, print!"
```

This noun-verb consistency in naming classes and methods aligns well with the way we naturally understand and communicate about objects and actions in the real world, contributing to the readability and comprehensibility of our code. Moreover, it lends itself well to the principle of encapsulation in OOP, where objects manage their own behavior (methods) and state (member variables).

Maintaining this convention allows developers to write code that's more intuitive, self-documenting, and easier to maintain. It creates a common language and understanding among developers, reducing the cognitive load when reading code and making the code base easier to navigate and reason about. Therefore, it's recommended to adhere to these conventions in OOP.

Naming variables

Variable names should reflect the data they hold. A good variable name describes the kind of value the variable contains, not just its purpose in the algorithm you've written.

Avoiding magic numbers, Numerical values with unexplained meanings in the source code. They can lead to code that is harder to read, understand, and maintain. Let's consider a `MessageSender` class that sends messages, and if a message size is greater than a certain limit, it splits the message into chunks:

```
class MessageSender {
public:
    void send_message(const std::string& message) {
        if (message.size() > 1024) {
            // Split the message into chunks and send
        } else {
            // Send the message
        }
    }
};
```

In the preceding code, `1024` is a magic number. It likely represents a maximum message size, but it's not immediately clear. It can confuse others (or future you) reading your code. Here's a refactored example with a named constant:

```
class MessageSender {
    constexpr size_t MAX_MESSAGE_SIZE = 1024;
public:
    void send_message(const std::string& message) {
        if (message.size() > MAX_MESSAGE_SIZE) {
            // Split the message into chunks and send
```

```
        } else {
            // Send the message
        }
    }
};
```

In this refactored version, we've replaced the magic number `1024` with a named constant, `MAX_MESSAGE_SIZE`. It's now clear that `1024` is the maximum message size. Using named constants in this way makes your code more readable and maintainable. If the maximum message size needs to change in the future, you only need to update it in one place.

Utilize namespaces

Namespaces in C++ are incredibly valuable in preventing naming conflicts and properly structuring your code. The issue of naming conflicts, or collisions, arises when two or more identifiers in a program bear the same name. For instance, you might have a class named `Id` in two subsystems of your application – networking representing a connection ID and user ID in user management. Using them both without namespaces would cause a naming collision, and the compiler wouldn't know which `Id` you are referring to in your code.

To mitigate this, C++ provides the `namespace` keyword to encapsulate a functionality under a unique name. Namespaces are designed to solve the problem of name conflicts. By wrapping your code inside a namespace, you prevent it from colliding with same-named identifiers in other portions of your code or third-party libraries.

Here's an example:

```
namespace product_name {
    class Router {
        // class implementation
    };
}

// To use it elsewhere in the code
product_name::Router myRouter;
```

In this case, `product_name::Router` won't conflict with any other `Router` class in your product's code or a third-party library. If you develop library code, it is highly recommended to wrap all its entities, such as classes, functions, and variables, in a namespace. This will prevent name clashes with other libraries or the user's code.

It's common in C++ to mirror the project's directory structure with namespaces, making it easier to understand where different parts of the code base are located. For example, if you have a file at the ProductRepo/Networking/Router.cpp path, you might declare the Router class like this:

```
namespace product_name {
    namespace networking {
        class Router {
            // class implementation
        };
    }
}
```

You can then refer to the class with the fully qualified name product_name::networking::Router.

However, it's worth noting that until C++20, the language didn't natively support a module system that could replace or enhance the functionality provided by namespaces. With the arrival of modules in C++20, some of the practices might be changing, but understanding namespaces and their usage in naming is still vital.

Another way to use namespaces is to express the complexity levels of your code. For example, library code may have entities expected to be used by library consumers and internal ones. The following code snippet demonstrates this approach:

```
// communication/client.hpp
namespace communication {
class Client {
public:
    // public high-level methods
private:
    using HttpClient = communication::advanced::HttpClient;
    HttpClient inner_client_;
};
} // namespace communication

// communication/http/client.hpp
namespace communication::advanced::http {
class Client {
    // Lower-level implementation
};
} // namespace communication::advanced
```

In this extended example, the `communication::Client` class provides a high-level interface for sending and receiving messages. It uses the `advanced::http::Client` class for the actual implementation, but this detail is hidden from the users of the library. They don't need to know about the advanced class unless they aren't satisfied with the functionality provided by the default client and need more control.

The `Client` class, in the `communication::http::advanced` namespace, provides more low-level functionality that gives users more control over the details of the communication.

This organization makes it clear what functionality is intended for most users (`Client`) and what is provided for more advanced usage (`HttpClient`). Using namespaces in this way also helps avoid name clashes and keeps the code base well-organized. This approach is used by many libraries and frameworks – for example, it's common for Boost libraries to have a `detail` namespace for internal implementation.

The use of domain-specific language

If there are well-established terms in the problem domain, use them in your code. This can make your code easier to understand for people familiar with the domain. For example, in finance, terms such as "portfolio," "asset," "bond," "equity," "ticker," and "dividend" are commonly used. If you're writing an application related to finance, it's beneficial to use these terms in your class and variable names, as they clearly convey their roles to anyone with a background in finance.

Consider the following code snippet:

```cpp
class Portfolio {
public:
    void add_asset(std::unique_ptr<Asset> asset) {
        // add the asset to the portfolio
    }

    double total_dividend() const {
        // calculate the total dividends of the portfolio
    }

private:
    std::vector<std::unique_ptr<Asset>> assets_;
};

using Ticker = std::string;

class Asset {
```

```cpp
public:
    Asset(const Ticker& ticker, int64_t quantity) :
        ticker_{ticker},
        quantity_{quantity} {}
    virtual Asset() = default;
    virtual double total_dividend() const = 0;
    auto& ticker() const { return ticker_; }
    int64_t quantity() const { return quantity_; }
private:
    Ticker ticker_;
    int64_t quantity_;
};

class Bond : public Asset {
public:
    Bond(const Ticker& ticker, int64_t quantity) :
        Asset{ticker, quantity} {}
    double total_dividend() const override {
        // calculate bond dividend
    }
};

class Equity : public Asset {
public:
    Equity(const Ticker& ticker, int64_t quantity) :
        Asset{ticker, quantity} {}
    double total_dividend() const override {
        // calculate equity dividend
    }
};
```

In this example, `Portfolio`, `Asset`, `Bond`, `Equity`, `Ticker`, and `total_dividend()` are all terms that are directly borrowed from the domain of finance. A developer or stakeholder who's familiar with finance will understand the purpose of these classes and methods just by their names. This helps to create a common language between the developers, stakeholders, and domain experts, which can greatly facilitate communication and understanding. Note that it is not recommended to use `double` in real-world financial applications, as it does not have an accurate enough representation to prevent rounding errors from accumulating when doing arithmetic with monetary values.

Remember, the goal of these principles is to make your code as clear and easy to understand as possible. Writing code is not just about communicating with the computer; it's also about communicating with other developers, including your future self.

Balancing long names and comments in code

Proper naming conventions play a critical role in the clarity and readability of your code. Names of classes, methods, and variables should be descriptive enough to convey their purpose and functionality. Ideally, a well-chosen name can replace the need for additional comments, making your code self-explanatory.

However, there's a delicate balance to be struck. While long, descriptive names can be helpful, excessively long names can also be cumbersome and detract from the readability of code. On the other hand, overly short names can be ambiguous and make the code harder to understand. The key is to find the right balance – names should be long enough to convey their purpose, but not so long as to be unwieldy.

Consider this example from a hypothetical networking application:

```
class Router {
public:
    void route(const Message& message, Id receiver) {
        auto message_content = message.get_content();
        // Code to route the 'message_content' to the appropriate
'receiver'
    }
private:
    // Router's private members
};
```

In this case, the `route` method name and the `message`, `receiver`, and `message_content` variable names are all sufficiently descriptive to understand what the method does and what each variable represents. Additional comments to explain their roles aren't necessary.

That being said, there are cases where language constructs can't fully express the intent or nuances of your code, such as when relying on specific behavior from a third-party library or when coding a complex algorithm. In these cases, additional comments are necessary to provide context or explain why certain decisions were made.

Take this, for instance:

```
void route(const Message& message, Id receiver) {
    auto message_content = message.get_content();

    // Note: The routing_library has an idiosyncratic behavior where
    // it treats receiver id as one-indexed. Hence we need to
increment by 1.
    receiver++;
    // Code to route the 'message_content' to the appropriate
'receiver'
}
```

In this case, the comment is necessary to highlight a specific behavior of the third-party routing library, which isn't immediately apparent from the language constructs alone.

As a general rule, strive to make your code as self-explanatory as possible through good naming practices, but don't hesitate to use comments when they're necessary to provide important context or clarify complex logic. Remember, the ultimate goal is to create code that is easy to read, understand, and maintain.

Exploring popular C++ coding conventions – Google, LLVM, and Mozilla

In the realm of C++ programming, adhering to a consistent coding convention is crucial to ensure code clarity and maintainability. Among the myriad of styles available, three prominent conventions stand out for their widespread use and distinct approaches – Google's C++ Style Guide, LLVM Coding Standards, and Mozilla's Coding Style. This overview delves into the key aspects of each, highlighting their unique practices and philosophies:

- **Google's C++ Style Guide**: Google's guidelines are designed for internal use, but they are widely adopted in the open source community. Key features include the following:

  - **Filenames**: Use `.cc` and `.h` extensions for implementation and header files, respectively

  - **Variable names**: Regular variables use lowercase with underscores, class members have a trailing underscore, and constants are in `kCamelCase`

  - **Class names**: Use `CamelCase` for class names

  - **Indentation and formatting**: Use spaces instead of tabs, with a two-space indent

  - **Pointer and reference expressions**: Place `*` or `&` with the variable name (`int* ptr`, not `int *ptr`)

  - **Limitations**: Avoid non-const global variables, and prefer algorithms over loops where possible

- **LLVM Coding Standards**: Used in the LLVM Compiler Infrastructure, these standards emphasize readability and efficiency:

  - **Filenames**: Source files use the `.cpp` extension, and header files use `.h`.

  - **Variable names**: Variables and functions use `camelBack` style. Member variables have a trailing underscore.

  - **Class names**: Classes and structs are in `CamelCase`.

  - **Indentation and formatting**: Two spaces for indentation, with a strong emphasis on readability and avoiding over-compact code.

- **Pointer and reference expressions**: Place * or & adjacent to the type (`int *ptr`, not `int* ptr`).

- **Modern C++ usage**: Encourages the use of modern C++ features and patterns.

- **Mozilla Coding Style**: While not as universally recognized as Google or LLVM, Mozilla's coding style is still significant, especially in projects related to their technologies:

  - **Filenames**: Uses the `.cpp` and `.h` extensions

  - **Variable names**: Use `camelCase` for variables and functions, `CamelCase` for classes, and `SCREAMING_SNAKE_CASE` for constants

  - **Class names**: `CamelCase` is used for class names

  - **Indentation and formatting**: Prefer four spaces for indentation, and follow a clear block separation style

  - **Pointer and reference expressions**: Similar to LLVM, place * or & adjacent to the type

  - **Emphasis on performance**: Encourages writing efficient code with a focus on browser performance

Each of these conventions has its own philosophy and rationale. Google's style guide emphasizes consistency within a vast code base and across a large number of developers. LLVM's standards focus on clean, efficient code that leverages modern C++ features. Mozilla's style balances readability and performance, reflecting its origins in web technology development. It's important to choose a style that aligns with your project's goals, team size, and the specific technologies you work with.

Summary

In this chapter, we explored the critical role of naming programming. We recognized that good, consistent naming practices elevate the readability and maintainability of code, while also aiding in its self-documentation.

We pondered over the balance between using long descriptive names and shorter names supplemented by comments, understanding that both have their place in different contexts. The use of domain-specific language in naming was recommended for clarity, while "magic numbers" were cautioned against due to their opacity.

The influence of a variable's scope and lifetime on its naming was also discussed, emphasizing the need for more descriptive names for those variables with larger scopes and longer lifetimes.

The chapter wrapped up by stressing the value of adhering to a coding convention for naming, which instills consistency across the code base, thereby streamlining the code reading and comprehension process.

The insights garnered from this chapter serve as a foundation for the upcoming discussion on effectively leveraging the rich static type system in C++ for safer, cleaner, and clearer code. In the next chapter, we will shift our focus to the effective utilization of C++'s rich static type system.

6

Utilizing a Rich Static Type System in C++

In modern software development, the notion of "type" has transcended its primitive definition, evolving into a rich and expressive language feature that encapsulates more than just data representation. In C++, a language renowned for its performance and flexibility, the static type system serves as a powerful tool, enabling developers to write code that's not only robust and efficient but also self-documenting and maintainable.

The significance of types in C++ extends beyond the mere categorization of data. By enforcing strict compile-time checks, the language's type system reduces runtime errors, improves readability, and fosters a more intuitive understanding of code. With the advent of modern C++ standards, the opportunities to leverage types have further expanded, bringing forth elegant solutions to common programming challenges.

However, these powerful features can often be underutilized. Primitive data types such as integers are frequently misused to represent concepts such as time durations, leading to code that lacks expressiveness and can be prone to errors. Pointers, although flexible, can lead to null dereferencing issues, making the code base fragile.

In this chapter, we'll explore the rich landscape of C++'s static type system, focusing on advanced and modern techniques that help to mitigate these problems. From using the `<chrono>` library to represent time durations to employing `not_null` wrappers and `std::optional` for safer pointer handling, we'll delve into practices that embody the essence of strong typing.

We'll also look at external libraries such as Boost, which offer additional utilities to enhance type safety. Throughout the chapter, real-world examples will illustrate how these tools and techniques can be seamlessly integrated into your code, empowering you to harness the full potential of C++'s type system.

By the end of this chapter, you'll gain a deep understanding of how to utilize types to write more robust, readable, and expressive code, tapping into the true power of C++.

Utilizing Chrono for time duration

One of the best examples of how C++'s type system can be leveraged to write more robust code is the <chrono> library. Introduced in C++11, this header provides a set of utilities to represent time durations and points in time, as well as perform time-related operations.

Managing time-related functions using plain integers or structures such as timespec can be a bug-prone approach, especially when dealing with different units of time. Imagine a function that takes an integer representing a timeout in seconds:

```
void wait_for_data(int timeout_seconds) {
    sleep(timeout_seconds); // Sleeps for timeout_seconds seconds
}
```

This approach lacks flexibility and can lead to confusion when handling various time units. For example, if a caller mistakenly passes milliseconds instead of seconds, it can cause unexpected behavior.

By contrast, using <chrono> to define the same function makes the code more robust and expressive:

```
#include <chrono>
#include <thread>

void wait_for_data(std::chrono::seconds timeout) {
    std::this_thread::sleep_for(timeout); // Sleeps for the specified
timeout
}
```

The caller can now pass the timeout using a strongly-typed duration, such as std::chrono::seconds(5), and the compiler ensures that the correct unit is used. Moreover, <chrono> provides seamless conversion between different time units, allowing the caller to specify the timeout in seconds, milliseconds, or any other unit, without ambiguity. The following snippet illustrates the usage with different units:

```
wait_for_data(std::chrono::milliseconds(150));
```

By embracing the strong typing offered by <chrono>, the code becomes clearer, more maintainable, and less susceptible to common bugs related to time representation.

Improving Pointer Safety with not_null and std::optional

In C++, pointers are a fundamental part of the language, allowing direct memory access and manipulation. However, the flexibility that pointers offer comes with certain risks and challenges. Here, we'll explore how modern C++ techniques can enhance pointer safety.

The pitfalls of raw pointers

Raw pointers, while powerful, can be a double-edged sword. They provide no information about the ownership of the object they point to, and they can easily become "dangling" pointers, pointing to memory that has been deallocated. Dereferencing a null or dangling pointer leads to undefined behavior, which can result in hard-to-diagnose bugs.

Using not_null from the Guidelines Support Library

The not_null wrapper provided by the **Guidelines Support Library** (**GSL**) aims to overcome the challenges associated with raw pointers. By using not_null, you can clearly signal that a pointer should never be null:

```
#include <gsl/gsl>
void process_data(gsl::not_null<int*> data) {
    // Data is guaranteed not to be null here
}
```

If a user passes a null pointer to this function as follows, the application will be terminated:

```
int main() {
    int* p = nullptr;
    process_data(p); // this will terminate the program
    return 0;
}
```

However, if the pointer is passed as process_data(nullptr), the application will fail in compile time:

```
source>: In function 'int main()':
<source>:9:16: error: use of deleted function 'gsl::not_null<T>::not_
null(std::nullptr_t) [with T = int*; std::nullptr_t = std::nullptr_t]'
    9 |       process_data(nullptr);
      |       ~~~~~~~~~~~~^~~~~~~~~
In file included from <source>:1:
/opt/compiler-explorer/libs/GSL/trunk/include/gsl/pointers:131:5:
note: declared here
  131 |       not_null(std::nullptr_t) = delete;
      |       ^~~~~~~~
```

This promotes robust code by catching potential null pointer errors early, thus reducing runtime errors.

Extending not_null to smart pointers

`gsl::not_null` is not limited to raw pointers; it can also be used with smart pointers such as `std::unique_ptr` and `std::shared_ptr`. This allows you to combine the benefits of modern memory management with the additional safety guarantees that `not_null` provides.

With std::unique_ptr

`std::unique_ptr` ensures that the ownership of a dynamically allocated object is unique, and it automatically deletes the object when it is no longer needed. By using `not_null` with `unique_ptr`, you can also ensure that the pointer is never null:

```
#include <gsl/gsl>
#include <memory>

void process_data(gsl::not_null<std::unique_ptr<int>> data) {
    // Data is guaranteed not to be null here
}

int main() {
    auto data = std::make_unique<int>(42);
    process_data(std::move(data)); // Safely passed to the function
}
```

With std::shared_ptr

Similarly, `gsl::not_null` can be used with `std::shared_ptr`, which enables shared ownership of an object. This allows you to write functions that accept shared pointers without having to worry about nullity:

```
#include <gsl/gsl>
#include <memory>

void process_data(gsl::not_null<std::shared_ptr<int>> data) {
    // Data is guaranteed not to be null here
}

int main() {
    auto data = std::make_shared<int>(42);
    process_data(data); // Safely passed to the function
}
```

These examples demonstrate how `not_null` can seamlessly integrate with modern C++ memory management techniques. By enforcing that a pointer (whether raw or smart) cannot be null, you further reduce the potential for runtime errors and make code more robust and expressive.

Utilizing std::optional for optional values

Sometimes, a pointer is used to indicate an optional value, where `nullptr` signifies the absence of a value. C++17 introduced `std::optional`, which provides a type-safe way to represent optional values:

```
#include <optional>
std::optional<int> fetch_data() {
    if (/* some condition */)
        return 42;
    else
        return std::nullopt;
}
```

Using `std::optional` provides clear semantics and avoids the pitfalls associated with using pointers for this purpose.

A comparison between raw pointers and nullptr

Both `not_null` and `std::optional` offer advantages over raw pointers. While raw pointers can be null or dangling, leading to undefined behavior, `not_null` prevents null pointer errors at compile time, and `std::optional` provides a clear way to represent optional values.

Consider the following example using raw pointers:

```
int* findValue() {
    // ...
    return nullptr; // No value found
}
```

This code might lead to confusion and bugs, especially if the caller forgets to check for `nullptr`. By using `not_null` and `std::optional`, you can make the code more expressive and less error-prone.

Leveraging std::expected for expected results and errors

While `std::optional` elegantly represents an optional value, sometimes you need to convey more information about why a value might be missing. In such cases, `std::expected` provides a way to return either a value or an error code, making code more expressive and the error handling more robust.

Consider a scenario where you have a function that retrieves a value from a network, and you want to handle network errors. You might define an enumeration for the various network errors:

```
enum class NetworkError {
    Timeout,
    ConnectionLost,
    UnknownError
};
```

You can then use `std::expected` to define a function that returns either an `int` value or `NetworkError`:

```cpp
#include <expected>
#include <iostream>

std::expected<int, NetworkError> fetch_data_from_network() {
    // Simulating network operation...
    if (/* network timeout */) {
        return std::unexpected(NetworkError::Timeout);
    }
    if (/* connection lost */) {
        return std::unexpected(NetworkError::ConnectionLost);
    }

    return 42; // Successfully retrieved value
}

int main() {
    auto result = fetch_data_from_network();
    if (result) {
        std::cout << "Value retrieved: " << *result << '\n';
    } else {
        std::cout << "Network error: ";
        switch(result.error()) {
            case NetworkError::Timeout:
                std::cout << "Timeout\n";
                break;
            case NetworkError::ConnectionLost:
                std::cout << "Connection Lost\n";
                break;
            case NetworkError::UnknownError:
                std::cout << "Unknown Error\n";
                break;
        }
    }
}
```

Here, `std::expected` captures both the successful case and various error scenarios, allowing for clear and type-safe error handling. This example illustrates how modern C++ types such as `std::expected` enhance expressiveness and safety, allowing you to write code that more accurately models complex operations.

By embracing these modern C++ tools, you can enhance pointer safety in your code, reducing bugs and making your intentions clear.

Strong typing with enum class and scoped enumerations

Strong typing is a cornerstone of robust, maintainable software, and C++ provides several mechanisms to facilitate it. Among these, enum class, introduced in C++11, is a particularly effective tool for creating strongly typed enumerations that can make your programs both more robust and easier to understand.

A review of enum class

Traditional enumerations in C++ suffer from a few limitations – they can implicitly convert to integers, potentially causing errors if misused, and their enumerators are introduced into the enclosing scope, leading to name clashes. enum class, also known as scoped enumerations, addresses these limitations:

```
// Traditional enum
enum ColorOld { RED, GREEN, BLUE };
int color = RED; // Implicit conversion to int

// Scoped enum (enum class)
enum class Color { Red, Green, Blue };
// int anotherColor = Color::Red; // Compilation error: no implicit
conversion
```

The benefits over traditional enums

Scoped enumerations offer several advantages:

- **Strong typing**: There are no implicit conversions between the enum class type and integers, ensuring that you can't accidentally misuse an enumerator as an integer

- **Scoped names**: Enumerators are scoped to enum class, reducing the likelihood of name collisions

- **An explicit underlying type**: enum class allows you to explicitly specify the underlying type, giving you precise control over the data representation:

    ```
    enum class StatusCode : uint8_t { Ok, Error, Pending };
    ```

The ability to specify the underlying type is particularly useful for serializing data to binary formats. It ensures that you have fine-grained control over how the data is represented at the byte level, facilitating easier data interchange with systems that may have specific binary format requirements.

Real-world scenarios

The advantages of `enum class` make it a powerful tool for various scenarios:

- **State machines**: When modeling system states, `enum class` provides a type-safe, expressive way to represent various possible states

- **Option sets**: Many functions have multiple behavior options, which can be neatly and safely encapsulated using scoped enumerations

- **The return status**: Functions that return status codes can benefit from the type-safety and scoping rules provided by `enum class`:

```cpp
enum class NetworkStatus { Connected, Disconnected, Error };

NetworkStatus check_connection() {
    // Implementation
}
```

By using `enum class` to create strongly typed, scoped enumerations, you can write code that is not only easier to understand but also less prone to errors. This feature represents another step forward in C++'s ongoing evolution toward a language that combines high performance with modern programming conveniences. Whether you're defining a complex state machine or simply trying to represent multiple options or statuses, `enum class` offers a robust, type-safe solution.

Leveraging the standard library's type utilities

Modern C++ offers a rich set of type utilities within the Standard library that enables developers to write more expressive, type-safe, and maintainable code. Two prominent examples are `std::variant` and `std::any`.

std::variant – a type-safe union

`std::variant` provides a type-safe way to represent a value that can be one of several possible types. Unlike a traditional `union`, which allows the programmer to treat the stored value as any of its member types, leading to potential undefined behavior, `std::variant` tracks the currently held type and ensures proper handling:

```cpp
#include <variant>
#include <iostream>

std::variant<int, double, std::string> value = 42;

// Using std::get with an index:
int intValue = std::get<0>(value); // Retrieves the int value
```

```
// Using std::get with a type:
try {
    double doubleValue = std::get<double>(value); // Throws std::bad_
variant_access
} catch (const std::bad_variant_access& e) {
    std::cerr << "Bad variant access: " << e.what() << '\n';
}

// Using std::holds_alternative:
if (std::holds_alternative<int>(value)) {
    std::cout << "Holding int\n";
} else {
    std::cout << "Not holding int\n";
}
```

The benefits over traditional unions

- **Type safety**: With traditional unions, it's up to the programmer to track the currently held type, and misuse can lead to undefined behavior. std::variant, conversely, keeps track of the current type and provides safe access through functions such as std::get and std::holds_alternative.

- **Automatic construction and destruction**: Unlike unions, std::variant automatically constructs and destroys the held object when you assign a new value, managing the object's lifetime correctly.

- **Exception handling**: When attempting to access a type not currently held by the variant using std::get, a std::bad_variant_access exception is thrown, making error handling more transparent and easier to manage.

- **Standard library integration**: std::variant can be used with standard library functions such as std::visit, providing elegant ways to handle various types.

std::any – type-safe containers for any type

std::any is a container that can hold any type but maintains type safety by requiring explicit casting to the correct type. This allows for flexible handling of data without sacrificing type integrity:

```
#include <any>
#include <iostream>
#include <stdexcept>

std::any value = 42;
```

```
try {
    std::cout << std::any_cast<int>(value); // Outputs 42
    std::cout << std::any_cast<double>(value); // Throws std::bad_any_
cast
} catch(const std::bad_any_cast& e) {
    std::cerr << "Bad any_cast: " << e.what();
}
```

The advantages of using `std::any` include the following:

- **Flexibility**: It can store any type, making it suitable for heterogeneous collections or flexible APIs

- **Type safety**: Requires explicit casting, preventing accidental misinterpretation of the contained value

- **Encapsulation**: Allows you to pass around values without exposing their concrete types, supporting more modular and maintainable code

Advanced type techniques

As you delve deeper into C++, you'll find that the language offers an array of advanced techniques for enhancing type safety, readability, and maintainability. In this section, we'll explore a few of these advanced concepts, providing practical examples for each.

Templates – specializing for type safety

Templates are a powerful feature in C++, but you may want to impose certain constraints or specializations based on types. One way to do this is via template specialization, which allows you to define custom behavior for certain types.

For example, let's say you have a generic function to find the maximum element in an array:

```
template <typename T>
T find_max(const std::vector<T>& arr) {
    // generic implementation
    return *std::max_element(arr.begin(), arr.end());
}
```

Now, let's say you want to provide a specialized implementation for `std::string` that is case-insensitive:

```
template <>
std::string find_max(const std::vector<std::string>& arr) {
    return *std::max_element(arr.begin(), arr.end(),
                             [](const std::string& a, const
std::string& b) {
```

```
                                             return strcasecmp(a.c_str(), b.c_
str()) < 0;
                                      });
}
```

With this specialized version, calls to `find_max` with `std::string` will use a case-insensitive comparison.

Creating custom type traits

Sometimes, the standard type traits may not be sufficient for your needs. You can create your own custom type traits to encapsulate type-based logic. For instance, you might want a type trait to identify whether a class has a specific member function:

```
template <typename T, typename = void>
struct has_custom_method : std::false_type {};

template <typename T>
struct has_custom_method<T, std::void_t<decltype(&T::customMethod)>> :
std::true_type {};
```

You can use this custom trait like any other type trait:

```
static_assert(has_custom_method<MyClass>::value, "MyClass must have a
customMethod");
```

Type aliases for readability and maintainability

Type aliases can improve the readability and maintainability of your code by providing meaningful names for complex types. For example, instead of writing out `std::unordered_map<std::string, std::vector<int>>` repeatedly, you could create a type alias:

```
using StringToIntVectorMap = std::unordered_map<std::string,
std::vector<int>>;
```

Now, you can use `StringToIntVectorMap` in your code, making it more readable and easier to maintain:

```
StringToIntVectorMap myMap;
```

Type aliases can also be templated, allowing for even more flexibility:

```
template <typename Value>
using StringToValueMap = std::unordered_map<std::string, Value>;
```

By employing these advanced type techniques, you add another layer of safety, readability, and maintainability to your C++ code. These methods give you more control over how types behave in templates, how they're checked, and how they're represented, ensuring that you can write code that's as robust as it is efficient.

Avoiding common pitfalls in advanced type usage

Writing robust code with type checks

Type-checking is one of the pillars that contributes to the robustness and safety of a program. While C++ is strongly typed, it does allow for some flexibility (or leniency, depending on your perspective) that can lead to errors if not carefully managed. Here are some techniques and best practices to write robust C++ code by leveraging type checks.

Using type traits for compile-time checks

The C++ Standard library offers a set of type traits in the `<type_traits>` header, which allows you to inspect and make decisions based on types at compile time. For example, if you have a generic function that should only accept unsigned integral types, you can enforce this using `static_assert`:

```
#include <type_traits>

template <typename T>
void foo(T value) {
    static_assert(std::is_unsigned<T>::value, "foo() requires an
unsigned integral type");
    // ... function body
}
```

Leveraging constexpr if

C++17 introduced `constexpr if`, enabling you to write conditional code that's evaluated at compile time. This can be very useful for type-specific operations in template code:

```
template <typename T>
void bar(T value) {
    if constexpr (std::is_floating_point<T>::value) {
        // Handle floating-point types
    } else if constexpr (std::is_integral<T>::value) {
        // Handle integral types
    }
}
```

Strong types for function arguments

C++ allows type aliases, which can sometimes make it difficult to understand the purpose of a function argument. For example, a function declared as void process(int, int); is not very informative. Is the first integer a length? Is the second one an index? One way to mitigate this is by using strong type definitions, such as the following:

```
struct Length { int value; };
struct Index { int value; };

void process(Length l, Index i);
```

Now, the function signature provides semantic meaning, making it less likely for the developer to swap the arguments accidentally.

Implicit conversions and type coercion

A case of accidental file creation

In C++ development, it's common to define classes with constructors that accept various argument types for flexibility. However, this comes with the risk of unintentional implicit conversions. To illustrate this point, consider the following code snippet involving a File class and a clean function:

```cpp
#include <iostream>

class File {
public:
    File(const std::string& path) : path_{path} {
        auto file = fopen(path_.c_str(), "w");
        // check if file is valid
        // handle errors, etc
        std::cout << "File ctor\n";
    }

    auto& path() const {
        return path_;
    }

    // other ctors, dtor, etc

private:
    FILE* file_ = nullptr;
    std::string path_;
};
```

```
void clean(const File& file) {
    std::cout << "Removing the file: " << file.path() << std::endl;
}

int main() {
    auto random_string = std::string{"blabla"};
    clean(random_string);
}
```

The output demonstrates the issue clearly:

```
File ctor
Removing the file: blabla
```

The compiler automatically converts std::string to a File object due to the absence of the explicit keyword in the constructor, thereby triggering an unintended side-effect – the creation of a new file.

The utility of explicit

To mitigate such risks, the explicit keyword can be employed. By marking a constructor as explicit, you instruct the compiler to disallow implicit conversions for that constructor. Here's how the corrected File class would look:

```
class File {
public:
    explicit File(const std::string& path) : path_{path} {
        auto file = fopen(path_.c_str(), "w");
        // check if file is valid
        // handle errors, etc
        std::cout << "File ctor\n";
    }

    // ... rest of the class
};
```

With this change, the clean(random_string); line would result in a compilation error, effectively preventing accidental file creation.

A light-hearted caveat

While our example might be somewhat simplified for educational purposes (yes, there's no need to roll your own File class – we have libraries for that!), it serves to underline a critical aspect of type safety in C++. A seemingly innocuous constructor can lead to unexpected behavior if not explicitly guarded against implicit conversions.

So, remember, when you're defining constructors, it pays to be explicit about your intentions. You never know when you might accidentally start a "file party" you never intended to host.

Summary

As we've traversed the vast landscape of C++'s rich static type system, it's worth taking a moment to reflect on how far we've come. From the earliest days of C++, where raw pointers and loosely typed arrays reigned supreme, to the modern era of `std::optional`, `std::variant`, and `enum class`, the language has evolved substantially in its approach to type safety.

The real power of these advances shines through when we consider how they improve not just individual code snippets but also entire software systems. Embracing C++'s robust type constructs can help us write safer, more readable, and ultimately, more maintainable code. Features such as the `std::optional` and `not_null` wrappers reduce the chance of null pointer errors. Advanced techniques such as template specialization and custom type traits offer unprecedented control over type behavior. These are not just academic exercises; they are practical tools for the everyday C++ programmer.

Looking ahead, the trajectory of C++ suggests an increasingly nuanced and powerful type system. As the language continues to evolve, who knows what innovative type-related features may lie on the horizon? Perhaps future versions of C++ will offer even more dynamic type checking, or maybe they'll introduce new constructs that we can't yet imagine.

In the next chapter, we'll pivot from the nitty-gritty of types to the grand architecture of classes, objects, and object-oriented programming in C++. While types give us the building blocks, it's these larger constructs that help us assemble those blocks into towering structures of sustainable software design. Until then, may your types be strong, your pointers never null, and your code forever robust.

7
Classes, Objects, and OOP in C++

In this chapter, we delve into the sophisticated realm of classes, objects, and **object-oriented programming** (**OOP**) in C++. Tailored for the advanced C++ practitioner, our focus will be on elevating your understanding of class design, method implementation, inheritance, and template usage, steering clear of introductory explanations of these concepts. Our goal is to enhance your ability to construct robust and efficient software architectures using advanced object-oriented techniques.

The discussion begins by examining the intricate considerations necessary when defining classes, guiding you through the decision-making process to determine the best candidates for class encapsulation. This includes distinguishing situations where a simpler data structure, such as a struct, might be more appropriate, thereby optimizing both performance and readability.

Further, we explore the design of methods within classes—highlighting various types of methods, such as accessors, mutators, and factory methods, and establishing conventions that promote code clarity and maintainability. Special attention is given to advanced method design practices, including const correctness and visibility scopes, which are pivotal for securing and optimizing access to class data.

Inheritance, a cornerstone of OOP, is scrutinized not only for its benefits but also its liabilities. To provide a balanced perspective, we present alternatives such as composition and interface segregation that might better serve your design goals in certain scenarios. This nuanced discussion aims to equip you with the discernment necessary to choose the best inheritance strategy or its alternatives, depending on the specific requirements and constraints of your projects.

Expanding the discussion to generic programming, we delve into sophisticated template usage, which includes advanced techniques such as template metaprogramming. This section aims to demonstrate how templates can be leveraged to create highly reusable and efficient code. Additionally, we will touch upon the design of APIs using OOP principles, emphasizing how well-crafted interfaces can significantly enhance the usability and longevity of software components.

Each topic is punctuated with practical examples and case studies drawn from real-world applications, illustrating how these advanced techniques are applied in modern software development. By the end of this chapter, you should possess a deeper understanding of how to utilize OOP features in C++ to craft elegant, efficient, and scalable software architectures.

Good candidates for classes

Identifying good candidates for classes in OOP involves looking for entities that naturally encapsulate both data and behavior.

Cohesion

A class should represent a set of functionalities that are tightly related to each other. This means all the methods and data in the class are directly related to the specific functionalities it provides. For example, a `Timer` class is a good candidate because it encapsulates properties and methods related to timing (start, stop, reset times), maintaining high cohesion.

Encapsulation

Entities that have attributes and behaviors that should be shielded from outside interference or misuse can be encapsulated in a class.

A `BankAccount` class encapsulates the balance (attribute) and behaviors such as `deposit`, `withdraw`, and `transfer`, ensuring that balance manipulations are done only through controlled and safe operations.

Reusability

Classes should be designed to be reused across different parts of a program or even in different programs.

A `DatabaseConnection` class that manages database connections can be reused in multiple applications that require database interactions, handling connection, disconnection, and error management.

Abstraction

A class should provide a simplified interface by hiding complex logic from the user, representing a higher level of abstraction. For example, the standard library has classes such as `std::vector` that abstract the complexities of dynamic arrays, providing a simple interface for array operations.

Real-world entities

Classes often represent objects from the real world that are relevant to the system being modeled.

In a flight reservation system, classes such as `Flight`, `Passenger`, and `Ticket` are good candidates because they directly represent real-world objects with clear attributes and behaviors.

Manage complexity

Classes should help in managing complexity by breaking down large problems into smaller, more manageable parts.

Here is an example – in graphic editing software, a `GraphicObject` class might serve as a base class for more specific graphic objects such as `Circle`, `Rectangle`, and `Polygon`, organizing graphic properties and functionalities systematically.

Minimizing class responsibilities through encapsulation

Encapsulation is a fundamental concept in OOP that involves bundling the data (attributes) and the methods (functions) that operate on the data into a single unit or class. It not only hides the internal state of the object from the outside but also modularizes its behavior, making the software easier to manage and extend. However, how much functionality and data a class should encapsulate can significantly affect the maintainability and scalability of your application.

Over-encapsulation in classes – a common pitfall

In practice, encapsulating too much functionality and data within a single class is a common mistake that can lead to several issues. This often results in a **god object** – a class that controls too many different parts of the application, doing too much work on its own. Such classes are typically hard to understand, difficult to maintain, and problematic to test.

Let's look at an example of a badly encapsulated `Car` class.

Consider the following example of a `Car` class that attempts to manage not only the car's basic properties but also detailed aspects of its internal systems such as the engine, transmission, and entertainment system:

```
#include <iostream>
#include <string>

class Car {
private:
    std::string _model;
    double _speed;
    double _fuel_level;
    int _gear;
    bool _entertainment_system_on;

public:
```

```cpp
    Car(const std::string& model) : _model(model), _speed(0), _fuel_
level(50), _gear(1), _entertainment_system_on(false) {}

    void accelerate() {
        if (_fuel_level > 0) {
            _speed += 10;
            _fuel_level -= 5;
            std::cout << "Accelerating. Current speed: " << _speed <<
" km/h, Fuel level: " << _fuel_level << " liters" << std::endl;
        } else {
            std::cout << "Not enough fuel." << std::endl;
        }
    }

    void change_gear(int new_gear) {
        _gear = new_gear;
        std::cout << "Gear changed to: " << _gear << std::endl;
    }

    void toggle_entertainment_system() {
        _entertainment_system_on = !_entertainment_system_on;
        std::cout << "Entertainment System is now " << (_
entertainment_system_on ? "on" : "off") << std::endl;
    }

    void refuel(double amount) {
        _fuel_level += amount;
        std::cout << "Refueling. Current fuel level: " << _fuel_level
<< " liters" << std::endl;
    }
};
```

This Car class is problematic because it tries to manage too many aspects of the car's functionality, which are better handled by specialized components.

Proper encapsulation using composition

A better approach is to use composition to delegate responsibilities to other classes, each handling a specific part of the system's functionality. This not only adheres to the Single Responsibility Principle but also makes the system more modular and easier to maintain.

Here is an example of a well-designed Car class using composition:

```cpp
#include <iostream>
#include <string>
```

```cpp
class Engine {
private:
    double _fuel_level;

public:
    Engine() : _fuel_level(50) {}

    void consume_fuel(double amount) {
        _fuel_level -= amount;
        std::cout << "Consuming fuel. Current fuel level: " << _fuel_
level << " liters" << std::endl;
    }

    void refuel(double amount) {
        _fuel_level += amount;
        std::cout << "Engine refueled. Current fuel level: " << _fuel_
level << " liters" << std::endl;
    }

    double get_fuel_level() const {
        return _fuel_level;
    }
};

class Transmission {
private:
    int _gear;

public:
    Transmission() : _gear(1) {}

    void change_gear(int new_gear) {
        _gear = new_gear;
        std::cout << "Transmission: Gear changed to " << _gear <<
std::endl;
    }
};

class EntertainmentSystem {
private:
    bool _is_on;

public:
    EntertainmentSystem() : _is_on(false) {}
```

```cpp
    void toggle() {
        _is_on = !_is_on;
        std::cout << "Entertainment System is now " << (_is_on ? "on"
: "off") << std::endl;
    }
};

class Car {
private:
    std::string _model;
    double _speed;
    Engine _engine;
    Transmission _transmission;
    EntertainmentSystem _entertainment_system;

public:
    Car(const std::string& model) : _model(model), _speed(0) {}

    void accelerate() {
        if (_engine.get_fuel_level() > 0) {
            _speed += 10;
            _engine.consume_f
uel(5);
            std::cout << "Car accelerating. Current speed: " << _speed
<< " km/h" << std::endl;
        } else {
            std::cout << "Not enough fuel to accelerate." <<
std::endl;
        }
    }

    void change_gear(int gear) {
        _transmission.change_gear(gear);
    }

    void toggle_entertainment_system() {
        _entertainment_system.toggle();
    }

    void refuel(double amount) {
        _engine.refuel(amount);
    }
};
```

In this refined design, the `Car` class acts as a coordinator among its components rather than directly managing every detail. Each subsystem – engine, transmission, and entertainment system – handles its own state and behavior, leading to a design that is easier to maintain, test, and extend. This example showcases how appropriate encapsulation and composition can significantly enhance the structure and quality of object-oriented software.

Usage of structs and classes in C++

In C++, both structs and classes are used to define user-defined types that can contain data and functions. The primary difference between them lies in their default access levels: members of a class are private by default, while members of a struct are public. This distinction subtly influences their typical uses in C++ programming.

Structs – ideal for passive data structures

Structs in C++ are particularly suited for creating passive data structures where the primary purpose is to store data without encapsulating too much behavior. Due to their public-by-default nature, structs are often used when you want to allow direct access to the data members, which can simplify code and reduce the need for additional functions to manipulate data.

The following list outlines the instances when you should use structs:

*   **Data objects**: Structs are perfect for creating **plain old data** (**POD**) structures. These are simple objects that primarily hold data and have little or no functionality (methods). For example, structs are often used to represent coordinates in space, RGB color values, or settings configurations where direct access to data fields is more convenient than going through getters and setters:

    ```cpp
    struct Color {
        int red = 0;
        int green = 0;
        int blue = 0;
    };

    struct Point {
        double x = 0.0;
        double y = 0.0;
        double z = 0.0;
    };
    Fortunately, C++ 11 and C++ 20 provide aggregate initialization
    and designated initializers, making it easier to initialize
    structs with default values.
    // C++ 11
        auto point = Point {1.1, 2.2, 3.3};
    // C++ 20
        auto point2 = Point {.x = 1.1, .y = 2.2, .z = 3.3};
    ```

If C++ 20 is not available for your project, you can utilize C99-designated initializers to achieve a similar effect:

```
auto point3 = Point {.x = 1.1, .y = 2.2, .z = 3.3};
```

- **Interoperability**: Structs are useful in interfacing with code in C or in systems where data alignment and layout are critical. They ensure compatibility and performance in low-level operations, such as hardware interfacing or network communication.

- **Lightweight containers**: When you need a lightweight container to group together a few variables, structs provide a more transparent and less cumbersome way than classes. They are ideal for small aggregations where encapsulation isn't a primary concern.

Classes – encapsulating complexity

Classes are the backbone of C++ OOP and are used to encapsulate data and behavior into a single entity. The private-by-default access specifier encourages the hiding of internal state and implementation details, promoting a more rigorous design that follows encapsulation and abstraction principles.

The following list explains when you should use classes:

- **Complex systems**: For components that involve complex data manipulations, state management, and interface control, classes are the preferred choice. They provide mechanisms for data protection and interface abstraction, which are crucial for maintaining the integrity and stability of software systems:

```
class Car {
private:
    int speed;
    double fuel_level;

public:
    void accelerate();
    void brake();
    void refuel(double amount);
};
```

- **Behavior encapsulation**: Classes are ideal when the functionality (methods) is as important as the data. Encapsulating behaviors with data into classes allows for more maintainable and error-free code, as operations on the data are tightly controlled and clearly defined.

- **Inheritance and polymorphism**: Classes support inheritance and polymorphism, enabling the creation of complex object hierarchies that can be extended and modified dynamically. This is essential in many software design patterns and advanced system architectures.

Choosing between a struct and a class in C++ should be guided by the intended use: structs for simple, transparent data containers where direct data access is acceptable or necessary, and classes for more complex systems where encapsulation, behavior, and interface control are required. Understanding and utilizing the strengths of each can lead to cleaner, more efficient, and scalable code.

Common method types in classes – getters and setters

In OOP, particularly in languages such as Java, **getters** and **setters** are standard methods that serve as the primary interface for accessing and modifying the private data members of a class. These methods provide controlled access to an object's properties, adhering to the encapsulation principle, which is a cornerstone of effective object-oriented design.

Purpose and conventions of getters and setters

Getters (also known as accessors) are methods used to retrieve the value of a private field. They do not modify the data. Setters (also known as mutators) are methods that allow the modification of private fields based on the input they receive. These methods enable the internal state of an object to remain consistent and valid by potentially enforcing constraints or conditions when data is set.

Here are the conventions of getters and setters:

- **Naming**: Typically, a getter for a property, x, is named `get_x()`, and the setter is named `set_x(value)`. This naming convention is almost universal in Java and is commonly adopted in other programming languages that support class-based OOP.

- **Return types and parameters**: A getter for a property returns the same type as the property itself and takes no parameters, whereas a setter returns void and takes a parameter of the same type as the property it sets.

Here is an example of this in C++:

```
class Person {
private:
    std::string _name;
    int _age;

public:
    // Getter for the name property
    std::string get_name() const { return _name; }

    // Setter for the name property
    void set_name(const std::string& name) { _name = name; }

    // Getter for the age property
    int get_age() const { return _age; }
```

```
    // Setter for the age property
    void set_age(int age) {
        if (age >= 0) { // validate the age
            _age = age;
        }
    }
};
```

Usefulness and recommendations

Controlled access and validation: Getters and setters encapsulate the fields of a class, providing controlled access and validation logic. This helps to maintain the integrity of the data, ensuring that no invalid or inappropriate values are set.

Flexibility: By using getters and setters, developers can change the underlying implementation of how the data is stored and retrieved without changing the external interface of the class. This can be particularly useful in maintaining backward compatibility or when the data representation needs to be changed for optimization.

Consistency: These methods can enforce rules that need to be maintained consistently throughout an object's life cycle. For example, ensuring that a field never holds a null value or adheres to a specific format.

When to use getters and setters, and when not to

The rule of thumb is to use getters and setters in classes where encapsulation, business logic, or inheritance complexities are present. For example, for the Car and Engine classes with relatively complex logic, getters and setters are essential to maintain the integrity of the data and ensure that the system functions correctly. On the other hand, for a simple data structure such as Point or Color, where the primary purpose is to hold data without much behavior, using a struct with public data members might be more appropriate. Note that if the struct is a part of a library or API, it might be beneficial to provide getters and setters for future extensibility.

This nuanced approach allows developers to balance between control and simplicity, choosing the most appropriate tool for the specific requirements of their software components.

Inheritance in C++

Inheritance and composition are two fundamental OOP concepts that enable the creation of complex and reusable software designs in C++. They facilitate code reuse and help in modeling real-world relationships, though they operate differently.

Inheritance allows one class, known as the derived or subclass, to inherit properties and behaviors from another class, the base or superclass. This enables the derived class to reuse the code in the base class while extending or overriding its functionality. For instance, consider a BaseSocket class and its derived classes, TcpSocket and UdpSocket. The derived classes inherit the basic functionality of BaseSocket and add their specific implementations:

```
class BaseSocket {
public:
    virtual ssize_t send(const std::vector<uint8_t>& data) = 0;
    virtual ~BaseSocket() = default;
};

class TcpSocket : public BaseSocket {
public:
    ssize_t send(const std::vector<uint8_t>& data) override {
        // Implement TCP-specific send logic here
    }
};

class UdpSocket : public BaseSocket {
public:
    ssize_t send(const std::vector<uint8_t>& data) override {
        // Implement UDP-specific send logic here
    }
};
```

In this example, the TcpSocket and UdpSocket classes inherit from BaseSocket, demonstrating how inheritance promotes code reuse and establishes an "is-a" relationship. Inheritance also supports polymorphism, allowing objects of the derived class to be treated as instances of the base class, enabling dynamic method binding.

Composition, on the other hand, involves creating classes by including objects of other classes. Instead of inheriting from a base class, a class is composed of one or more objects from other classes, which are used to achieve the desired functionality. This represents a "has-a" relationship. For example, consider a CommunicationChannel class that can own BaseSocket. The CommunicationChannel class uses the BaseSocket object to implement its communication functionality, demonstrating composition:

```
class CommunicationChannel {
public:
    CommunicationChannel(std::unique_ptr<BaseSocket> sock) : _
socket(sock) {}

    bool transmit(const std::vector<uint8_t>& data) {
```

```cpp
        size_t total_sent = 0;
        size_t data_size = data.size();

        while (total_sent < data_size) {
            ssize_t bytesSent = _socket->send({data.begin() + total_
sent, data.end()}});
            if (bytesSent < 0) {
                std::cerr << "Error sending data." << std::endl;
                return false;
            }
            total_sent += bytesSent;
        }

        std::cout << "Communication channel transmitted " << total_
sent << " bytes." << std::endl;
        return true;
    }
private:
    std::unique_ptr<BaseSocket> _socket;

};

int main() {
    TcpSocket tcp;
    CommunicationChannel channel(std::make_unique<TcpSocket>());
    std::vector<uint8_t> data = {1, 2, 3, 4, 5};

    if (channel.transmit(data)) {
        std::cout << "Data transmitted successfully." << std::endl;
    } else {
        std::cerr << "Data transmission failed." << std::endl;
    }

    return 0;
}
```

In this example, the CommunicationChannel class contains a BaseSocket object and uses it to implement its functionality. The transmit method sends data in chunks until all data is sent, checking for errors (when the return value is less than 0). This demonstrates how composition offers flexibility, allowing objects to be dynamically assembled at runtime. It also promotes better encapsulation by containing objects and exposing only necessary interfaces, thereby avoiding tight coupling between classes and making the code more modular and easier to maintain.

In summary, both inheritance and composition are essential tools in C++ for creating reusable and maintainable code. Inheritance is suitable for scenarios with a clear hierarchical relationship and where polymorphism is needed, while composition is ideal for assembling complex behaviors from simpler components, offering flexibility and better encapsulation. Understanding when to use each approach is key to effective object-oriented design.

Evolution of inheritance in C++

Originally, inheritance was seen as a powerful tool for reducing code duplication and enhancing the expressiveness of code. It allowed for the creation of a derived class that inherits properties and behavior from a base class. However, as the use of C++ grew in complex systems, the limitations of inheritance as a one-size-fits-all solution became apparent.

Implementation of inheritance at the binary level

Interestingly, on a binary level, inheritance in C++ is implemented similarly to composition. Essentially, the derived class contains an instance of the base class within its structure. This can be visualized in a simplified ASCII diagram:

```
+-------------------+
|   Derived Class   |
|-------------------|
|  Base Class Part  | <- Base class subobject
|-------------------|
| Derived Class Data| <- Additional data members of the derived class
+-------------------+
```

In this layout, the base class part of the derived class object contains all the data members that belong to the base class, and directly after it in memory, the additional data members of the derived class are placed. Note that the actual order of data members in memory can be influenced by factors such as alignment requirements and compiler optimizations.

Pros and cons of inheritance

Here are the pros of inheritance:

- **Code reuse**: Inheritance allows developers to create a new class based on an existing class, making it easy to reuse code and reduce redundancy. Let's use an example from a media player system to demonstrate inheritance and code reuse in a different context. We'll design a class hierarchy for various types of media content that a player might handle, such as audio files, video files, and podcasts.

The MediaContent class will serve as the base class for all types of media content. It will encapsulate common attributes and behaviors such as title, duration, and basic playback controls (play, pause, stop):

```cpp
#include <iostream>
#include <string>

// Base class for all media content
class MediaContent {
protected:
    std::string _title;
    int _duration; // Duration in seconds

public:
    MediaContent(const std::string& title, int duration)
        : _title(title), _duration(duration) {}

    auto title() const { return _title; }
    auto duration() const { return duration; }

    virtual void play() = 0; // Start playing the content
    virtual void pause() = 0;
    virtual void stop() = 0;

    virtual ~MediaContent() = default;
};
```

The Audio class extends MediaContent, adding specific attributes related to audio files, such as bitrate:

```cpp
class Audio : public MediaContent {
private:
    int _bitrate; // Bitrate in kbps

public:
    Audio(const std::string& title, int duration, int bitrate)
        : MediaContent(title, duration), _bitrate(bitrate) {}

    auto bitrate() const { return _bitrate; }

    void play() override {
        std::cout << "Playing audio: " << title << ", Duration: " << duration
                    << "s, Bitrate: " << bitrate << "kbps" << std::endl;
    }
```

```
    void pause() override {
        std::cout << "Audio paused: " << title << std::endl;
    }

    void stop() override {
        std::cout << "Audio stopped: " << title << std::endl;
    }
};
```

Similarly, the Video class extends MediaContent and introduces additional attributes such as resolution:

```
class Video : public MediaContent {
private:
    std::string _resolution; // Resolution as width x height

public:
    Video(const std::string& title, int duration, const
std::string& resolution)
        : MediaContent(title, duration), _resolution(resolution)
{}

    auto resolution() const { return _resolution; }

    void play() override {
        std::cout << "Playing video: " << title << ", Duration:
" << duration
                  << "s, Resolution: " << resolution <<
std::endl;
    }

    void pause() override {
        std::cout << "Video paused: " << title << std::endl;
    }

    void stop() override {
        std::cout << "Video stopped: " << title << std::endl;
    }
};
```

Here's how these classes could be used in a simple media player system:

```
int main() {
    Audio my_song("Song Example", 300, 320);
    Video my_movie("Movie Example", 7200, "1920x1080");
```

```
    my_song.play();
    my_song.pause();
    my_song.stop();

    my_movie.play();
    my_movie.pause();
    my_movie.stop();

    return 0;
}
```

In this example, both `Audio` and `Video` inherit from `MediaContent`. This allows us to reuse the `title` and `duration` attributes and requires the implementation of the playback controls (`play`, `pause`, `stop`) tailored to each media type. This hierarchy demonstrates how inheritance facilitates code reuse and system extensibility while enabling specific behaviors for different types of media content in a unified framework. Each class adds only what is unique to its type, adhering to the principle that base classes provide common functionality and derived classes extend or modify that functionality for specific needs.

- **Polymorphism**: Through inheritance, C++ supports polymorphism, which allows for the use of a base class reference to refer to an object of a derived class. This enables dynamic method binding and a flexible interface to multiple derived types. Our media content hierarchy can be used for implementing a media player that can handle different types of media content uniformly:

```cpp
class MediaPlayer {
private:
    std::vector<std::unique_ptr<MediaContent>> _playlist;

public:
    void add_media(std::unique_ptr<MediaContent> media) {
        _playlist.push_back(std::move(media));
    }

    void play_all() {
        for (auto& media : _playlist) {
            media->play();
            // Additional controls can be implemented
        }
    }
};

int main() {
    MediaPlayer player;
    player.add(std::make_unique<Audio>("Jazz in Paris", 192,
320));
```

```
        player.add(std::make_unique<Video>("Tour of Paris", 1200,
    "1280x720"));

        player.play_all();

        return 0;
    }
```

The `add` method accepts media content of any type that derives from `MediaContent`, demonstrating polymorphism by using a base class pointer to refer to derived class objects. This is enabled by storing the media items in `std::vector` of `std::unique_ptr<MediaContent>`. The `play_all` method iterates through the stored media and calls the play method on each item. Despite the actual media type being different (audio or video), the media player treats them all as `MediaContent`. The correct play method (from either `Audio` or `Video`) is invoked at runtime, which is an example of dynamic polymorphism (also known as dynamic dispatch).

- **Hierarchical structuring**: It provides a natural way to organize related classes in a hierarchical manner that models real-world relationships.

Here is the con of inheritance:

- **Tight coupling**: Inheritance creates a tight coupling between base and derived classes. Changes to the base class can inadvertently affect derived classes, leading to fragile code that can break when base classes are modified. The following example illustrates the issue of tight coupling through inheritance in a software system. We'll use a scenario involving an online store that manages different types of discounts using a class hierarchy.

Base class – Discount

The `Discount` class provides the basic structure and functionality for all types of discounts. It calculates a discount based on a percentage reduction;

```cpp
#include <iostream>

class Discount {
protected:
    double _discount_percent;  // Percent of discount

public:
    Discount(double percent) : _discount_percent(percent) {}

    virtual double apply_discount(double amount) {
        return amount * (1 - _discount_percent / 100);
    }
};
```

Derived class – SeasonalDiscount

The `SeasonalDiscount` class extends `Discount` and modifies the discount calculation based on seasonal factors, such as increasing the discount during the holiday season:

```
class SeasonalDiscount : public Discount {
public:
    SeasonalDiscount(double percent) : Discount(percent) {}

    double apply_discount(double amount) override {
        // Let's assume the discount increases by an additional 5%
during holidays
        double additional = 0.05;  // 5% extra during holidays
        return amount * (1 - (_discount_percent / 100 + additional));
    }
};
```

Derived class – ClearanceDiscount

The `ClearanceDiscount` class also extends `Discount`, designed for items on clearance where the discount might be significantly higher:

```
class ClearanceDiscount : public Discount {
public:
    ClearanceDiscount(double percent) : Discount(percent) {}

    double apply_discount(double amount) override {
        // Clearance items get an extra 10% off beyond the configured
discount
        double additional = 0.10;  // 10% extra for clearance items
        return amount * (1 - (_discount_percent / 100 + additional));
    }
};
```

Demonstration and tight coupling issue:

```
int main() {
    Discount regular(20); // 20% regular discount
    SeasonalDiscount holiday(20); // 20% holiday discount, plus extra
    ClearanceDiscount clearance(20); // 20% clearance discount, plus
extra

    std::cout << "Regular Price $100 after discount: $" << regular.
apply_discount(100) << std::endl;
    std::cout << "Holiday Price $100 after discount: $" << holiday.
apply_discount(100) << std::endl;
```

```
    std::cout << "Clearance Price $100 after discount: $" <<
clearance.apply_discount(100) << std::endl;

    return 0;
}
```

Tight coupling problems

The following is a list of tight-coupling problems:

- **Dependency on the base class method**: All subclasses are tightly coupled to the base class's method structure (`apply_discount`). Any change in the base class method's signature or the logic within `apply_discount` could necessitate changes in all derived classes.

- **Assumptions on internal logic**: Subclasses assume they can simply add to `_discount_percent`. If the formula in the base class changes (say, incorporating minimum or maximum caps), all subclasses might need extensive modifications to conform to the new logic.

- **Inflexibility**: The coupling makes it hard to modify the behavior of one discount type without risking impacts on others. This design lacks flexibility where independent evolution of discount calculation strategies might be necessary.

Solution – decouple with the strategy pattern

One way to reduce this coupling is to use the **strategy pattern**, which involves defining a family of algorithms (discount strategies), encapsulating each one, and making them interchangeable. This allows the discount algorithms to vary independently from the clients that use them:

```cpp
class DiscountStrategy {
public:
    virtual double calculate(double amount) = 0;
    virtual ~DiscountStrategy() {}
};

class RegularDiscountStrategy : public DiscountStrategy {
public:
    double calculate(double amount) override {
        return amount * 0.80; // 20% discount
    }
};

class HolidayDiscountStrategy : public DiscountStrategy {
public:
    double calculate(double amount) override {
        return amount * 0.75; // 25% discount
```

```
        }
};

class ClearanceDiscountStrategy : public DiscountStrategy {
public:
    double calculate(double amount) override {
        return amount * 0.70; // 30% discount
    }
};

// Use these strategies in a Discount context class
class Discount {
private:
    std::unique_ptr<DiscountStrategy> _strategy;

public:
    Discount(std::unique_ptr<DiscountStrategy> strat) : _
strategy(std::move(strat)) {}
    double apply_discount(double amount) {
        return _strategy->calculate(amount);
    }
};
```

This approach decouples the discount calculation from the client (Discount) using it, allowing each discount strategy to evolve independently without affecting others. A couple of others ways to reduce the coupling are:

- **Complexity**: Deep and complex inheritance hierarchies, along with the use of multiple inheritance, introduce a range of challenges that can complicate software design, making systems harder to understand, maintain, and evolve. When classes derive from multiple levels of base classes, forming extended chains of dependencies, understanding and modifying such classes demands an awareness of the entire inheritance chain. This depth increases complexity as changes in top-level classes can unpredictably affect functionality across all subclasses, leading to what is often referred to as "fragility" in software design.

  Multiple inheritance, where a class inherits characteristics from more than one base class, introduces its own set of problems. This approach can lead to the infamous "diamond problem," where ambiguities arise if both parent classes derive from a common ancestor and provide their own implementations of the same method. While languages such as C++ provide mechanisms such as virtual inheritance to address such issues, these solutions add layers of complexity and can introduce inefficiencies in memory management and method resolution.

Combining multiple inheritance with multi-level inheritance hierarchies, sometimes seen in more complex or "exotic" system designs, compounds these difficulties. For example, a class such as `HybridFlyingElectricCar` that inherits from both `ElectricCar` and `FlyingCar`, with each of these classes further inheriting from their respective hierarchies, results in a highly tangled class structure. This complexity makes the system tough to debug, extend, or reliably use, while also multiplying the challenges in testing and maintaining consistent behavior across various scenarios.

To manage the complications introduced by extensive use of inheritance, several strategies can be recommended. Favoring composition over inheritance often provides greater flexibility, allowing systems to be composed of well-defined, loosely coupled components rather than relying on rigid inheritance structures. Keeping inheritance chains short and manageable – generally no deeper than two or three levels – helps preserve system clarity and maintainability. Employing interfaces, particularly in languages such as Java and C#, offers a way to achieve polymorphic behavior without the overhead associated with inheritance. When multiple inheritance is unavoidable, it's crucial to ensure clear documentation and consider the use of interface-like structures or mixins, which can help minimize complexity and enhance system robustness.

- **Liskov Substitution Principle (LSP)**: We mentioned this principle earlier in this book; LSP states that objects of a superclass should be replaceable with objects of its subclasses without altering the desirable properties of the program (correctness, task performed, etc.). Inheritance can sometimes lead to violations of this principle, especially when subclasses diverge from the behavior expected by the base class. The following sections include typical problems related to violations of the LSP, illustrated with simple examples.

Unexpected behaviors in derived classes

When derived classes override methods of the base class in ways that change the expected behavior significantly, it can lead to unexpected results when these objects are used interchangeably:

```cpp
class Bird {
public:
    virtual void fly() {
        std::cout << "This bird flies" << std::endl;
    }
};

class Ostrich : public Bird {
public:
    void fly() override {
        throw std::logic_error("Ostriches can't fly!");
    }
};
```

```
void make_bird_fly(Bird& b) {
    b.fly();  // Expecting all birds to fly
}
```

Here, replacing a `Bird` object with an `Ostrich` object in the `make_bird_fly` function leads to a runtime error because ostriches can't fly, violating LSP. Users of the `Bird` class expect any subclass to fly, and `Ostrich` breaks this expectation.

Issues with method preconditions

If a derived class imposes stricter preconditions on a method than those imposed by the base class, it can limit the usability of the subclass and violate LSP:

```
class Payment {
public:
    virtual void pay(int amount) {
        if (amount <= 0) {
            throw std::invalid_argument("Amount must be positive");
        }
        std::cout << "Paying " << amount << std::endl;
    }
};

class CreditPayment : public Payment {
public:
    void pay(int amount) override {
        if (amount < 100) {  // Stricter precondition than the base
class
            throw std::invalid_argument("Minimum amount for credit
payment is 100");
        }
        std::cout << "Paying " << amount << " with credit" <<
std::endl;
    }
};
```

Here, the `CreditPayment` class cannot be used in place of `Payment` without potentially throwing an error for amounts below 100, even though such amounts are perfectly valid for the base class.

Solutions to LSP violations

- **Design with LSP in mind**: When designing your class hierarchy, ensure that any subclass can be used in place of a parent class without altering the desirable properties of the program

- **Use composition instead of inheritance**: If it doesn't make sense for the subclass to fully adhere to the base class's contract, use composition instead of inheritance

- **Clearly define behavioral contracts**: Document and enforce the expected behavior of base classes, and ensure that all derived classes adhere strictly to these contracts without introducing tighter preconditions or altering postconditions

By paying close attention to these principles and potential pitfalls, developers can create more robust and maintainable object-oriented designs.

While inheritance remains a valuable feature in C++, understanding when and how to use it effectively is crucial. The implementation detail that inheritance is akin to composition at the binary level highlights that it is fundamentally about structuring and accessing data within an object's memory layout. Practitioners must carefully consider whether inheritance or composition (or a combination of both) will best serve their design goals, especially regarding system flexibility, maintainability, and the robust application of OOP principles such as LSP. As with many features in software development, the key lies in using the right tool for the right job.

Templates and generic programming

Templates and generic programming are pivotal features of C++ that enable the creation of flexible and reusable components. While this chapter offers an overview of these powerful tools, it's important to note that the topic of templates, particularly template metaprogramming, is vast enough to fill entire books. For those seeking an in-depth exploration, dedicated resources on C++ templates and metaprogramming are recommended.

What are templates good for?

Templates are particularly useful in scenarios where similar operations need to be performed on different types of data. They allow you to write a single piece of code that works with any type. The following subsections outline some common use cases with examples.

Generic algorithms

Algorithms can operate on different types without rewriting the code for each type. For instance, the standard library's `std::sort` function can sort elements of any type as long as the elements can be compared:

```
#include <algorithm>
#include <vector>
#include <iostream>

template <typename T>
void print(const std::vector<T>& vec) {
    for (const T& elem : vec) {
        std::cout << elem << " ";
```

```
    }
    std::cout << std::endl;
}

int main() {
    std::vector<int> int_vec = {3, 1, 4, 1, 5};
    std::sort(int_vec.begin(), int_vec.end());
    print(int_vec); // Outputs: 1 1 3 4 5

    std::vector<std::string> string_vec = {"banana", "apple",
"cherry"};
    std::sort(string_vec.begin(), string_vec.end());
    print(string_vec); // Outputs: apple banana cherry

    return 0;
}
```

Container classes

Templates are heavily used in the standard library for containers such as std::vector, std::list, and std::map, which can store elements of any type:

```
#include <vector>
#include <iostream>

int main() {
    std::vector<int> int_vec = {1, 2, 3};
    std::vector<std::string> string_vec = {"hello", "world"};

    for (int val : int_vec) {
        std::cout << val << " ";
    }
    std::cout << std::endl;

    for (const std::string& str : string_vec) {
        std::cout << str << " ";
    }
    std::cout << std::endl;

    return 0;
}
```

Without the usage of templates, developer's options in using collections would be limited to creating separate classes for each type of collection (e.g., `IntVector`, `StringVector`, etc.), or demanding the use of a common base class, which would require type casting and lose type safety, for example:

```
class BaseObject {};

class Vector {
public:
    void push_back(BaseObject* obj);
};
```

Another option is to store some `void` pointers and cast them to the desired type when retrieving them, but this approach is even more error prone.

The standard library uses templates for smart pointers such as `std::unique_ptr` and `std::shared_ptr`, which manage the lifetime of dynamically allocated objects:

```
#include <memory>
#include <iostream>

int main() {
    std::unique_ptr<int> ptr = std::make_unique<int>(42);
    std::cout << "Value: " << *ptr << std::endl; // Outputs: Value: 42

    std::shared_ptr<int> shared_ptr = std::make_shared<int>(100);
    std::cout << "Shared Value: " << *shared_ptr << std::endl; //
Outputs: Shared Value: 100

    return 0;
}
```

Templates ensure type safety by allowing the compiler to check types during template instantiation, reducing runtime errors:

```
template <typename T>
T add(T a, T b) {
    return a + b;
}

int main() {
    std::cout << add<int>(5, 3) << std::endl;       // Outputs: 8
    std::cout << add<double>(2.5, 3.5) << std::endl; // Outputs: 6.0
    return 0;
}
```

How templates work

Templates in C++ are not actual code but serve as blueprints for code generation. When a template is instantiated with a specific type, the compiler generates a concrete instance of the template with the specified type replacing the template parameters.

Function templates

A function template defines a pattern for a function that can operate on different data types:

```cpp
template <typename T>
T add(T a, T b) {
    return a + b;
}

int main() {
    std::cout << add<int>(5, 3) << std::endl;       // Outputs: 8
    std::cout << add<double>(2.5, 3.5) << std::endl; // Outputs: 6.0
    return 0;
}
```

The actual generated functions after template instantiation would be something like this (depending on the compiler):

```cpp
int addInt(int a, int b) {
    return a + b;
}

double addDouble(double a, double b) {
    return a + b;
}
```

Class templates

A class template defines a pattern for a class that can operate on different data types:

```cpp
template <typename T>
class Box {
private:
    T content;
public:
    void set_content(const T& value) {
        content = value;
    }
```

```
    T get_content() const {
        return content;
    }
};

int main() {
    Box<int> intBox;
    intBox.set_content(123);
    std::cout << intBox.get_content() << std::endl; // Outputs: 123

    Box<std::string> stringBox;
    stringBox.set_content("Hello Templates!");
    std::cout << stringBox.get_content() << std::endl; // Outputs:
Hello Templates!
    return 0;
}
```

The actual generated classes after template instantiation would be something like this (depending on the compiler):

```
class BoxInt { /*Box<int>*/ };
class BoxString { /*Box<int>*/ };
```

How templates are instantiated

When a template is used with a specific type, the compiler creates a new instance of the template with the specified type. This process is known as **template instantiation** and can happen implicitly or explicitly:

- **Implicit instantiation**: This occurs when the compiler encounters a use of the template with specific types:

```
int main() {
    std::cout << add(5, 3) << std::endl; // The compiler infers
the type as int
    return 0;
}
```

- **Explicit instantiation**: The programmer specifies the type explicitly:

```
int main() {
    std::cout << add<int>(5, 3) << std::endl; // Explicitly
specifies the type as int
    return 0;
}
```

A real-world example of template usage in C++

In the realm of financial software, handling various types of assets and currencies in a flexible, type-safe, and efficient manner is crucial. C++ templates offer a powerful mechanism to achieve this flexibility by allowing developers to write generic and reusable code that can operate with any data type.

Imagine developing a financial system that must handle multiple currencies such as USD and EUR, and manage various assets such as stocks or bonds. By using templates, we can define classes that operate generically on these types without duplicating code for each specific currency or asset type. This approach not only reduces redundancy but also enhances the system's scalability and maintainability.

In the following sections, we will look at a detailed example of a financial system implemented using C++ templates. This example will show you how to define and manipulate prices in different currencies, how to create and manage assets, and how to ensure that operations remain type safe and efficient. Through this example, we aim to illustrate the practical benefits of using templates in real-world C++ applications and how they can lead to cleaner, more maintainable, and more robust code.

Defining currencies

When designing a financial system, it's essential to handle multiple currencies in a way that prevents errors and ensures type safety. Let's start by defining the requirements and exploring various design options.

Here are the requirements:

- **Type safety**: Ensure that different currencies cannot be mixed inadvertently
- **Scalability**: Easily add new currencies without significant code duplication
- **Flexibility**: Support various operations such as addition and subtraction on prices in a type-safe manner

Here are the design options:

- **Using primitive types (int, double)**: One approach is to represent currencies using primitive types such as `int` or `double`. However, this method has significant drawbacks. It allows for the accidental mixing of different currencies, leading to incorrect calculations:

```
double usd = 100.0;
double eur = 90.0;
double total = usd + eur; // Incorrectly adds USD and EUR
```

This approach is error prone and lacks type safety. Please note that using `double` for currency values is generally discouraged due to precision issues in floating-point arithmetic.

- **Inheritance from a base currency class**: Another approach is to define a base `Currency` class and inherit specific currencies from it. While this approach introduces some structure, it still allows for the mixing of different currencies and requires significant effort to implement each new currency:

```cpp
class Currency {
public:
    virtual std::string name() const = 0;
    virtual ~Currency() = default;
};

class USD : public Currency {
public:
    std::string name() const override { return "USD"; }
};

class Euro : public Currency {
public:
    std::string name() const override { return "EUR"; }
};

// USD and Euro can still be mixed inadvertently
```

- **Using templates for currency definition**: The most robust solution is to use templates to define currencies. This approach ensures type safety by preventing the mixing of different currencies at compile time. Each currency is defined as a `struct`, and operations are implemented using templates:

```cpp
struct Usd {
    static const std::string &name() {
        static std::string name = "USD";
        return name;
    }
};

struct Euro {
    static const std::string &name() {
        static std::string name = "EUR";
        return name;
    }
};

template <typename Currency>
class Price {
```

```cpp
public:
    Price(int64_t amount) : _amount(amount) {}
    int64_t count() const { return _amount; }

private:
    int64_t _amount;
};

template <typename Currency>
std::ostream &operator<<(std::ostream &os, const Price<Currency>
&price) {
    os << price.count() << " " << Currency::name();
    return os;
}

template <typename Currency>
Price<Currency> operator+(const Price<Currency> &lhs, const
Price<Currency> &rhs) {
    return Price<Currency>(lhs.count() + rhs.count());
}

template <typename Currency>
Price<Currency> operator-(const Price<Currency> &lhs, const
Price<Currency> &rhs) {
    return Price<Currency>(lhs.count() - rhs.count());
}

// User can define other arithmetic operations as needed
```

This template-based approach ensures that prices in different currencies cannot be mixed:

```cpp
int main() {
    Price<Usd> usd(100);
    Price<Euro> euro(90);

    // The following line would cause a compile-time error
    // source>:113:27: error: no match for 'operator+' (operand
types are 'Price<Usd>' and 'Price<Euro>')
    // Price<Usd> total= usd + euro;

    Price<Usd> total = usd+ Price<Usd>(50); // Correct usage
    std::cout << total<< std::endl; // Outputs: 150 USD

    return 0;
}
```

Defining assets

Next, we define assets that can be priced in different currencies. Using templates, we can ensure that each asset is associated with the correct currency:

```cpp
template <typename TickerT>
class Asset;

struct Apple {
    static const std::string &name() {
        static std::string name = "AAPL";
        return name;
    }

    static const std::string &exchange() {
        static std::string exchange = "NASDAQ";
        return exchange;
    }

    using Asset = class Asset<Apple>;
    using Currency = Usd;
};

struct Mercedes {
    static const std::string &name() {
        static std::string name = "MGB";
        return name;
    }

    static const std::string &exchange() {
        static std::string exchange = "FRA";
        return exchange;
    }

    using Asset = class Asset<Mercedes>;
    using Currency = Euro;
};

template <typename TickerT>
class Asset {
public:
    using Ticker   = TickerT;
    using Currency = typename Ticker::Currency;
```

```cpp
    Asset(int64_t amount, Price<Currency> price)
        : _amount(amount), _price(price) {}

    auto amount() const { return _amount; }
    auto price() const { return _price; }

private:
    int64_t _amount;
    Price<Currency> _price;
};

template <typename TickerT>
std::ostream &operator<<(std::ostream &os, const Asset<TickerT>
&asset) {
    os << TickerT::name() << ", amount: " << asset.amount()
        << ", price: " << asset.price();
    return os;
}
```

Using the financial system

Finally, we demonstrate how to use the defined templates to manage assets and prices:

```cpp
int main() {
    Price<Usd> usd_price(100);
    usd_price = usd_price + Price<Usd>(1);
    std::cout << usd_price << std::endl; // Outputs: 101 USD

    Asset<Apple> apple{10, Price<Usd>(100)};
    Asset<Mercedes> mercedes{5, Price<Euro>(100)};

    std::cout << apple << std::endl; // Outputs: AAPL, amount: 10,
price: 100 USD
    std::cout << mercedes << std::endl; // Outputs: MGB, amount: 5,
price: 100 EUR

    return 0;
}
```

Disadvantages of using templates in system design

While templates in C++ offer a powerful and flexible way to create type-safe, generic components, there are several disadvantages to this approach. These disadvantages are particularly relevant in the context of a financial system that deals with multiple currencies and assets. Understanding these potential drawbacks is essential when deciding to use templates in your design.

Code bloat

Templates can lead to code bloat, which is the increase in binary size due to the generation of multiple template instantiations. The compiler generates a separate version of the template code for each unique type instantiation. In a financial system that supports various currencies and assets, this can lead to a significant increase in the size of the compiled binary.

For example, if we have templates for `Price` and `Asset` instantiated with different types such as `Usd`, `Euro`, `Apple`, and `Mercedes`, the compiler generates separate code for each combination:

```
Price<Usd> usdPrice(100);
Price<Euro> euroPrice(90);
Asset<Apple> appleAsset(10, Price<Usd>(100));
Asset<Mercedes> mercedesAsset(5, Price<Euro>(100));
```

Each of these instantiations results in additional code, contributing to the overall binary size. As the number of supported currencies and assets grows, the impact of code bloat becomes more pronounced. Binary size can affect application performance, memory usage, and load times, especially in resource-constrained environments mostly due to lower cache efficiency.

Increased compilation times

Templates can significantly increase the compilation time of a project. Each instantiation of a template with a new type results in the generation of new code by the compiler. In a financial system that supports hundreds of currencies and assets from various countries and stock exchanges, the compiler will have to instantiate all the needed combinations, leading to longer build times.

For instance, say our system supports the following:

- 50 different currencies
- 10000 different asset types from various stock exchanges

Then, the compiler will need to generate code for each `Price` and `Asset` combination, resulting in a substantial number of template instantiations. This can considerably slow down the compilation process, affecting the development workflow, and less efficient feedback loop.

Less obvious interaction with the rest of the code

Template code can be complex and less obvious in terms of how it interacts with the rest of the code base. Developers who are less experienced with templates may find it challenging to understand and maintain template-heavy code. The syntax can be verbose, and compiler error messages can be difficult to decipher, making debugging and troubleshooting more complicated.

For example, a simple mistake in template parameters can lead to confusing error messages:

```
template <typename T>
class Price {
    // Implementation
};


Price<int> price(100); // Intended to be Price<Usd> but mistakenly
used int
```

In this case, the developer must understand templates and the specific error messages generated by the compiler to resolve the issue. This can be a barrier for less experienced developers.

C++ 20 provides concepts to improve template error messages and constraints, which can help make template code more readable and easier to understand. We can create a base class called `BaseCurrency` and derive all currency classes from it. This way, we can ensure that all currency classes have a common interface and can be used interchangeably:

```
struct BaseCurrency {
};


struct Usd : public BaseCurrency {
    static const std::string &name() {
        static std::string name = "USD";
        return name;
    }
};



// Define a concept for currency classes
template<class T, class U>
concept Derived = std::is_base_of<U, T>::value;


// Make sure that template parameter is derived from BaseCurrency
template <Derived<BaseCurrency> CurrencyT>
class Price {
public:
    Price(int64_t amount) : _amount(amount) {}
    int64_t count() const { return _amount; }

private:
    int64_t _amount;
};
```

After these changes, the attempt to instantiate `Price<int>` will result in a compile-time error, making it clear that the type must be derived from `BaseCurrency`:

```
In function 'int main()':
error: template constraint failure for 'template<class
CurrencyT> requires  Derived<CurrencyT, Currency> class Price'
 auto p = Price<int>(100);
          ^

note: constraints not satisfied
In substitution of 'template<class
CurrencyT> requires  Derived<CurrencyT, Currency> class Price [with
CurrencyT = int]':
```

C++ versions prior to C++ 20 also provide a way to prevent unintended template instantiations by using a combination of `std::enable_if` and `std::is_base_of` to enforce constraints on template parameters:

```
template <typename CurrencyT,
          typename Unused=typename std::enable_if<std::is_base_
of<BaseCurrency,CurrencyT>::value>::type>
class Price {
public:
    Price(int64_t amount) : _amount(amount) {}
    int64_t count() const { return _amount; }

private:
    int64_t _amount;
};
```

The attempt to initialize `Price<int>` will now result in a compile-time error, indicating that the type must be derived from `BaseCurrency`, however, the error message will be a bit cryptic:

```
error: no type named 'type' in 'struct std::enable_if<false, void>'
auto p = Price<int>(100);
       |                  ^
error: template argument 2 is invalid
```

Limited tool support and debugging

Debugging template code can be challenging due to limited tool support. Many debuggers do not handle template instantiations well, making it difficult to step through template code and inspect template parameters and instantiations. This can hinder the debugging process and make it harder to identify and fix issues.

For example, examining the state of a templated `Price<Usd>` object in a debugger might not provide clear insights into the underlying type and values, especially if the debugger does not fully support template parameter inspection.

Most autocomplete and IDE tools do not work very well with templates, because it is impossible for them to assume the type of the template parameter. This can make it harder to navigate and understand template-heavy code bases.

Advanced features of templates might be hard to use

Templates in C++ provide a mechanism for writing generic and reusable code. However, there are situations where the default template behavior needs to be customized for specific types. This is where template specialization comes into play. Template specialization allows you to define a special behavior for a specific type, ensuring that the template behaves correctly for that type.

Why use template specialization?

Template specialization is used when the general template implementation does not work correctly or efficiently for a particular type, or when a specific type requires a completely different implementation. This can happen due to various reasons, such as performance optimizations, special handling of certain data types, or compliance with specific requirements.

For example, consider a scenario where you have a general `Printer` template class that prints objects of any type. However, for `std::string`, you might want to add quotes around the string when printing it.

Basic template specialization example

Here is an example of how template specialization works:

```cpp
#include <iostream>
#include <string>

// General template
template <typename T>
class Printer {
public:
    void print(const T& value) {
        std::cout << value << std::endl;
    }
};

// Template specialization for std::string
template <>
class Printer<std::string> {
```

```
public:
    void print(const std::string& value) {
        std::cout << "\"" << value << "\"" << std::endl;
    }
};

int main() {
    Printer<int> int_printer;
    int_printer.print(123); // Outputs: 123

    Printer<std::string> string_printer;
    string_printer.print("Hello, World!"); // Outputs: "Hello, World!"
with quotes

    return 0;
}
```

In this example, the general `Printer` template class prints any type. However, for `std::string`, the specialized version adds quotes around the string when printing it.

Including the specialization header

When using template specialization, it is crucial to include the header file that contains the specialization definition. If the specialization header is not included, the compiler will instantiate the default version of the template, leading to incorrect behavior.

For example, consider the following files:

`printer.h` (General template definition):

```
#ifndef PRINTER_H
#define PRINTER_H

#include <iostream>

template <typename T>
class Printer {
public:
    void print(const T& value) {
        std::cout << value << std::endl;
    }
};

#endif // PRINTER_H
```

`printer_string.h` (Template specialization for `std::string`):

```
#ifndef PRINTER_STRING_H
#define PRINTER_STRING_H

#include "printer.h"
#include <string>

template <>
class Printer<std::string> {
public:
    void print(const std::string& value) {
        std::cout << "\"" << value << "\"" << std::endl;
    }
};

#endif // PRINTER_STRING_H
```

`main.cpp` (Using the template and specialization):

```
#include "printer.h"
// #include "printer_string.h" // Uncomment this line to use the
specialization

int main() {
    Printer<int> int_printer;
    int_printer.print(123); // Outputs: 123

    Printer<std::string> string_printer;
    string_printer.print("Hello, World!"); // Outputs: Hello, World!
without quotes if the header is not included

    return 0;
}
```

In this setup, if the `printer_string.h` header is not included in `main.cpp`, the compiler will use the default `Printer` template for `std::string`, resulting in incorrect behavior (printing the string without quotes).

Templates are a crucial part of the C++ programming language, offering powerful capabilities for creating generic, reusable, and type-safe code. They are indispensable in various scenarios, such as developing generic algorithms, container classes, smart pointers, and other utilities that need to work seamlessly with multiple data types. Templates enable developers to write flexible and efficient code, ensuring that the same functionality can be applied to different types without duplication.

However, the power of templates does not come without cost. The use of templates can lead to increased compilation times and code bloat, especially in systems that support a wide range of types and combinations. The syntax and resulting error messages can be complex and difficult to understand, posing a challenge for less experienced developers. Additionally, debugging template-heavy code can be cumbersome due to limited tool support and the intricate nature of template instantiations.

Moreover, templates can introduce less obvious interactions with the rest of the code base, which might cause issues if not managed properly. Developers must also be aware of advanced features, such as template specialization, which require careful inclusion of specialized headers to avoid incorrect behavior.

Given these caveats, it is essential for developers to think carefully before incorporating templates into their projects. While they provide significant benefits, the potential drawbacks necessitate a thoughtful approach to ensure that the advantages outweigh the complexities. Proper understanding and judicious use of templates can lead to more robust, maintainable, and efficient C++ applications.

Summary

In this chapter, we explored the intricacies of advanced C++ programming, focusing on class design, inheritance, and templates. We began with the principles of effective class design, emphasizing the importance of encapsulating the minimum necessary functionality and data to achieve better modularity and maintainability. Through practical examples, we highlighted both good and bad design practices. Moving on to inheritance, we examined its benefits, such as code reuse, hierarchical structuring, and polymorphism, while also addressing its drawbacks, including tight coupling, complex hierarchies, and potential violations of the LSP. We provided guidance on when to use inheritance and when to consider alternatives such as composition. In the section on templates, we delved into their role in enabling generic programming, allowing for flexible and reusable components that work with any data type. We discussed the advantages of templates, such as code reusability, type safety, and performance optimization, but also pointed out their disadvantages, including increased compilation times, code bloat, and the complexity of understanding and debugging template-heavy code. Throughout these discussions, we underscored the need for careful consideration and understanding when utilizing these powerful features to ensure robust and maintainable C++ applications. In the next chapter, we will shift our focus to API design, exploring best practices for creating clear, efficient, and user-friendly interfaces in C++.

8

Designing and Developing APIs in C++

In the world of software development, the design of **application programming interfaces (APIs)** is of paramount importance. Good APIs serve as the backbone of software libraries, facilitating interaction between different software components and enabling developers to leverage functionality efficiently and effectively. Well-designed APIs are intuitive, easy to use, and maintainable, playing a crucial role in the success and longevity of software projects. In this chapter, we will delve into principles and practices for designing maintainable APIs for libraries developed in C++. We will explore key aspects of API design, including clarity, consistency, and extensibility, and provide concrete examples to illustrate best practices. By understanding and applying these principles, you will be able to create APIs that not only meet the immediate needs of users but also remain robust and adaptable over time, ensuring that your libraries are both powerful and user-friendly.

Principles of minimalistic API design

Minimalistic APIs are designed to provide only the essential functionalities required to perform specific tasks, avoiding unnecessary features and complexity. The primary goal is to offer a clean, efficient, and user-friendly interface that facilitates easy integration and usage. Key benefits of minimalistic APIs include the following:

- **Ease of use**: Users can quickly understand and utilize the API without extensive learning or documentation, promoting faster development cycles

- **Maintainability**: Simplified APIs are easier to maintain, allowing for straightforward updates and bug fixes without introducing new complexities

- **Performance**: Leaner APIs tend to have better performance due to reduced overhead and more efficient execution paths

- **Reliability**: With fewer components and interactions, the likelihood of bugs and unexpected issues is minimized, leading to more reliable and stable software

Simplicity and clarity are fundamental principles in the design of minimalistic APIs. These principles ensure that the API remains accessible and user-friendly, enhancing the overall developer experience. Key aspects of simplicity and clarity include the following:

- **Straightforward interfaces**: Designing simple and clear interfaces helps developers quickly grasp the available functionalities, making it easier to integrate and use the API effectively

- **Reduced cognitive load**: By minimizing the mental effort required to understand and use the API, developers are less likely to make mistakes, leading to more efficient development processes

- **Intuitive design**: An API that adheres to simplicity and clarity aligns closely with common usage patterns and developer expectations, making it more intuitive and easier to adopt

Overengineering and unnecessary complexity can severely undermine the effectiveness of an API. To avoid these pitfalls, consider the following strategies:

- **Focus on core functionality**: Concentrate on delivering essential features that address the primary use cases. Avoid adding extraneous features that are not directly aligned with the core purpose of the API.

- **Iterative design**: Begin with a **minimum viable product** (**MVP**) and incrementally add features based on user feedback and actual needs rather than speculative requirements.

- **Clear documentation**: Provide comprehensive yet concise documentation that focuses on core functionality and common use cases. This helps prevent confusion and misuse.

- **Consistent naming conventions**: Use consistent and descriptive names for functions, classes, and parameters to enhance clarity and predictability.

- **Minimal dependencies**: Reduce the number of external dependencies to simplify the integration process and minimize potential compatibility issues.

Techniques for achieving minimalism

Functional decomposition is the process of breaking down complex functionalities into smaller, more manageable units. This technique is crucial for creating minimalistic APIs as it promotes simplicity and modularity. By decomposing functions, you ensure that each part of the API has a clear, well-defined purpose, which enhances maintainability and usability.

Key aspects of functional decomposition include the following:

- **Modular design**: Design the API such that each module or function handles a specific aspect of the overall functionality. This **separation of concerns** (**SoC**) ensures that each part of the API has a clear, well-defined purpose.

- **Single Responsibility Principle** (**SRP**): Each function or class should have one, and only one, reason to change. This principle helps in keeping the API simple and focused.

- **Reusable components**: By decomposing functions into smaller units, you can create reusable components that can be combined in different ways to achieve various tasks, enhancing the flexibility and reusability of the API.

Interface segregation aims to keep interfaces lean and focused on specific tasks, avoiding the design of monolithic interfaces that try to cover too many use cases. This principle ensures that clients only need to know about the methods that are relevant to them, making the API easier to use and understand.

Key aspects of interface segregation include the following:

- **Specific interfaces**: Instead of one large, general-purpose interface, design multiple smaller, specific interfaces. Each interface should cater to a specific aspect of the functionality.

- **User-centric design**: Consider the needs of the end users of your API. Design interfaces that are intuitive and provide only methods they need for their tasks, avoiding unnecessary complexity.

- **Reduced client impact**: Smaller, focused interfaces minimize the impact on clients when changes are necessary. Clients using a specific interface are less likely to be affected by changes in unrelated functionality.

Let us consider an example where a complex API class is responsible for various features, such as loading, processing, and saving data:

```
class ComplexAPI {
public:
    void initialize();
    void load_data_from_file(const std::string& filePath);
    void load_data_from_database(const std::string& connection_
string);
    void process_data(int mode);
    void save_data_to_file(const std::string& filePath);
    void save_data_to_database(const std::string& connection_string);
    void cleanup();
};
```

The major issue is that the class has too many responsibilities, mixing different data sources and sinks, leading to complexity and lack of focus. Let us start with extracting the loading and processing functionalities into separate classes:

```
class FileDataLoader {
public:
    explicit FileDataLoader(const std::string& filePath) :
filePath(filePath) {}
    void load() {
        // Code to load data from a file
    }
```

```
private:
    std::string filePath;
};

class DatabaseDataLoader {
public:
    explicit DatabaseDataLoader(const std::string& connection_string)
: _connection_string(connection_string) {}
    void load() {
        // Code to load data from a database
    }
private:
    std::string _connection_string;
};

class DataProcessor {
public:
    void process(int mode) {
        // Code to process data based on the mode
    }
};
```

The next step is to extract the saving functionalities into separate classes:

```
class DataSaver {
public:
    virtual void save() = 0;
    virtual ~DataSaver() = default;
};

class FileDataSaver : public DataSaver {
public:
    explicit FileDataSaver(const std::string& filePath) :
filePath(filePath) {}
    void save() override {
        // Code to save data to a file
    }
private:
    std::string filePath;
};

class DatabaseDataSaver : public DataSaver {
public:
    explicit DatabaseDataSaver(const std::string& connection_string) :
_connection_string(connection_string) {}
```

```
    void save() override {
        // Code to save data to a database
    }
private:
    std::string _connection_string;
};
```

Minimizing the number of dependencies required by the API is crucial for achieving minimalism. Fewer dependencies lead to a more stable, reliable, and maintainable API. Dependencies can complicate integration, increase the risk of compatibility issues, and make the API harder to understand.

Key strategies for reducing dependencies include the following:

- **Core functionality focus**: Concentrate on implementing core functionalities within the API itself, avoiding reliance on external libraries or components unless absolutely necessary.

- **Selective use of libraries**: When external libraries are required, choose those that are stable, well maintained, and widely used. Ensure that they align closely with the needs of your API.

- **Decoupled design**: Design the API in a way that it can function independently of external components as much as possible. Use **dependency injection** (**DI**) or other design patterns to decouple the implementation from specific dependencies.

- **Version management**: Carefully manage and specify versions of any dependencies to avoid compatibility issues. Ensure that updates to dependencies do not break the API or introduce instability.

Real-world examples of minimalistic API design

To solidify our understanding of these concepts, we will examine a few real-world examples of API design in C++. These examples will highlight common challenges and effective solutions, demonstrating how to apply the principles of good API design in practical scenarios. Through these examples, we aim to provide clear, actionable insights that you can apply to your own projects, ensuring that your APIs are not only functional but also elegant and maintainable. Let's dive into the intricacies of real-world API design and see how these principles come to life in practice:

- **JSON for Modern C++ (nlohmann/json)**: This library is an excellent example of minimalistic API design. It provides intuitive and straightforward methods for parsing, serializing, and manipulating JSON data in C++ and has the following benefits:

  - **Simplicity**: Clear and concise interface that is easy to use.

  - **Functional decomposition**: Each function handles a specific task related to JSON processing.

- **Minimal dependencies**: Designed to work with the C++ Standard Library, avoiding unnecessary external dependencies:

```
#include <nlohmann/json.hpp>

nlohmann::json j = {
    {"pi", 3.141},
    {"happy", true},
    {"name", "Niels"},
    {"nothing", nullptr},
    {"answer", {
        {"everything", 42}
    }},
    {"list", {1, 0, 2}},
    {"object", {
        {"currency", "USD"},
        {"value", 42.99}
    }}
};
```

- **SQLite C++ Interface (SQLiteCpp)**: This library offers a minimalistic interface for interacting with SQLite databases in C++. It has the following benefits:

  - **Simplicity**: Provides a straightforward and clear API for database operations.

  - **Interface segregation**: Separate classes for different database operations such as queries and transactions.

  - **Minimal dependencies**: Built to use SQLite and the C++ Standard Library:

```
#include <SQLiteCpp/SQLiteCpp.h>

SQLite::Database db("test.db", SQLite::OPEN_
READWRITE|SQLite::OPEN_CREATE);

db.exec("CREATE TABLE test (id INTEGER PRIMARY KEY, value
TEXT)");

SQLite::Statement query(db, "INSERT INTO test (value) VALUES
(?)");
query.bind(1, "Sample value");
query.exec();
```

Common pitfalls and how to avoid them

Overcomplication occurs when the API design includes unnecessary features or complexity, making it difficult to use and maintain. Here's how to mitigate this:

- **Avoidance strategy**: Focus on core functionalities required by the end users. Regularly review the API design to eliminate any unnecessary features.

Feature creep happens when additional features are continually added to the API, leading to increased complexity and reduced usability. Here's how you can avoid this:

- **Avoidance strategy**: Implement a strict feature prioritization process. Ensure that new features are aligned with the core purpose of the API and are necessary for the target users.

Important caveats of developing shared libraries in C++

Developing shared libraries in C++ requires careful consideration to ensure compatibility, stability, and usability. Originally, shared libraries were intended to promote code reuse, modularity, and efficient memory usage, allowing multiple programs to use the same library code simultaneously. This approach was expected to reduce redundancy, save system resources, and provide the ability to replace only parts of applications. While this approach worked well for widely used libraries, such as `libc`, `libstdc++`, OpenSSL, and others, it proved to be less efficient for applications. Shared libraries provided with an application can rarely be spotlessly replaced with a newer version. Usually, it is required to replace the whole installation kit, which includes the application and all its dependencies.

Nowadays, shared libraries are often used to enable interoperability between different programming languages. For instance, a C++ library might be used in applications written in Java or Python. This cross-language functionality extends the usability and reach of the library but introduces certain complexities and caveats that developers must consider.

Shared libraries within a single project

If the shared library is designed to be used within a single project and loaded by an executable compiled with the same compiler, then shared objects (or DLLs) with C++ interfaces are generally acceptable. However, this approach comes with caveats, such as the use of singletons, which can lead to issues with multithreading and unexpected initialization order. When singletons are used, managing their initialization and destruction in a multithreaded environment can be challenging, leading to potential race conditions and unpredictable behavior. Additionally, ensuring the correct order of initialization and destruction of global state is complex, which can result in subtle and hard-to-diagnose bugs.

Shared libraries for wider distribution

If the shared library is expected to be distributed more widely, where the developers cannot predict the compiler used by the end users or if the library might be used from other programming languages, then C++ shared libraries are not an ideal choice. This is primarily because the C++ **Application Binary Interface** (**ABI**) is not stable across different compilers or even different versions of the same compiler. C++ ABI instability arises from the complexity of the language, evolving standards, platform-specific variations, compiler optimizations, and differences in runtime and standard library implementations. This instability can lead to binary compatibility issues, making it challenging to distribute shared libraries and increasing maintenance overhead. For instance, C++ templates and inline functions are instantiated in every translation unit where they are used, leading to potential differences across compilers or compiler versions. Changes in C++ standards and compiler updates can also alter the ABI. Different operating systems and hardware platforms may have their own calling conventions and binary formats, adding to the complexity. Compiler optimizations can vary, and the implementation of the C++ Standard Library differs across compilers and platforms, further contributing to ABI incompatibilities. These issues can cause crashes or undefined behavior when modules compiled with different compilers or versions interact, necessitating careful versioning and management of libraries. One effective way to mitigate ABI instability is to use C interfaces, as the C ABI is much more stable. Despite not being formally standardized, the C ABI serves as a de facto standard glue between programming languages since all languages essentially need to support C interfaces to communicate with `libc` or operating system syscalls, which are also in C. A common solution to this problem is to develop a C wrapper around the C++ code and ship the C interface.

Example – MessageSender class

Next is an example demonstrating this approach, where we create a C++ `MessageSender` class and provide a C wrapper for it. The class has a constructor that initializes a `MessageSender` instance with a specified receiver and two overloaded `send` methods that allow sending messages either as a `std::vector<uint8_t>` instance or as a raw pointer with a specified length. The implementation prints messages to the console to demonstrate functionality.

Here's the C++ library implementation:

```
// MessageSender.hpp
#pragma once
#include <string>
#include <vector>

class MessageSender {
public:
    MessageSender(const std::string& receiver);
    void send(const std::vector<uint8_t>& message) const;
    void send(const uint8_t* message, size_t length) const;
```

```
};

// MessageSender.cpp
#include "MessageSender.h"
#include <iostream>

MessageSender::MessageSender(const std::string& receiver) {
    std::cout << "MessageSender created for receiver: " << receiver <<
std::endl;
}

void MessageSender::send(const std::vector<uint8_t>& message) const {
    std::cout << "Sending message of size: " << message.size() <<
std::endl;
}

void MessageSender::send(const uint8_t* message, size_t length) const
{
    std::cout << "Sending message of length: " << length << std::endl;
}
```

Here's the C wrapper implementation:

```
// MessageSender.h (C Wrapper Header)
#ifdef __cplusplus
extern "C" {
#endif

typedef void* MessageSenderHandle;

MessageSenderHandle create_message_sender(const char* receiver);
void destroy_message_sender(MessageSenderHandle handle);
void send_message(MessageSenderHandle handle, const uint8_t* message,
size_t length);

#ifdef __cplusplus
}
#endif

// MessageSenderC.cpp (C Wrapper Implementation)
#include "MessageSenderC.h"
#include "MessageSender.hpp"

MessageSenderHandle create_message_sender(const char* receiver) {
    return new(std::nothrow) MessageSender(receiver);
```

```
}

void destroy_message_sender(MessageSenderHandle handle) {
    MessageSender* instance = reinterpret_
cast<MessageSender*>(handle);
    assert(instance);
    delete instance;
}

void send_message(MessageSenderHandle handle, const uint8_t* message,
size_t length) {
    MessageSender* instance = reinterpret_
cast<MessageSender*>(handle);
    assert(instance);
    instance->send(message, length);
}
```

In this example, the C++ `MessageSender` class is defined in the `MessageSender.hpp` and `MessageSender.cpp` files. The class has a constructor that initializes a `MessageSender` instance with a specified receiver and two overloaded `send` methods that allow sending messages either as a `std::vector<uint8_t>` instance or as a raw pointer with a specified length. The implementation prints messages to the console to demonstrate functionality.

To make this C++ class usable from other programming languages or with different compilers, we create a C wrapper. The C wrapper is defined in the `MessageSender.h` and `MessageSenderC.cpp` files. The header file uses an `extern "C"` block to ensure that the C++ functions are callable from C, preventing name mangling. The C wrapper uses an opaque handle, `void*` (typedef as `MessageSenderHandle`), to represent the `MessageSender` instance in C, abstracting the actual C++ class.

The `create_message_sender` function allocates and initializes a `MessageSender` instance and returns a handle to it. Note that it uses `new(std::nothrow)` to avoid throwing exceptions in case of memory allocation failure. C or any other programming language that does not support exceptions can still use this function without issues.

The `destroy_message_sender` function deallocates the `MessageSender` instance to ensure proper cleanup. The `send_message` function calls the corresponding `send` method on the `MessageSender` instance using the handle, facilitating message sending.

By handling memory allocation and deallocation within the same binary, this approach avoids issues related to different memory allocators being used by the end user, which can lead to memory corruption or leaks. The C wrapper provides a stable and consistent interface that can be used across different compilers and languages, ensuring greater compatibility and stability. This method addresses the complexities of developing shared libraries and ensures their broad usability and reliability.

If the C++ library is expected to throw exceptions, it is important to handle them properly in C wrapper functions to prevent exceptions from propagating to the caller. For example, we can have the following exception types:

```
class ConnectionError : public std::runtime_error {
public:
    ConnectionError(const std::string& message) : std::runtime_
error(message) {}
};

class SendError : public std::runtime_error {
public:
    SendError(const std::string& message) : std::runtime_
error(message) {}
};
```

Then, the C wrapper functions can catch these exceptions and return appropriate error codes or messages to the caller:

```
// MessageSender.h (C Wrapper Header)
typedef enum {
    OK,
    CONNECTION_ERROR,
    SEND_ERROR,
} MessageSenderStatus;
// MessageSenderC.cpp (C Wrapper Implementation)
MessageSenderStatus send_message(MessageSenderHandle handle, const
uint8_t* message, size_t length) {
    try {
        MessageSender* instance = reinterpret_
cast<MessageSender*>(handle);
        instance->send(message, length);
        return OK;
    } catch (const ConnectionError&) {
        return CONNECTION_ERROR;
    } catch (const SendError&) {
        return SEND_ERROR;
    } catch (...) {
        std::abort();
    }
}
```

Note that we use std::abort in case of an unknown exception, as it is not safe to propagate unknown exceptions across language boundaries.

This example illustrates how to create a C wrapper around a C++ library to ensure compatibility and stability when developing shared libraries. By following these guidelines, developers can create robust, maintainable, and widely compatible shared libraries, ensuring their usability across various platforms and programming environments.

Summary

In this chapter, we explored critical aspects of designing and developing shared libraries in C++. Shared libraries were initially conceived to promote code reuse, modularity, and efficient memory usage by allowing multiple programs to utilize the same library code simultaneously. This approach reduces redundancy and conserves system resources.

We delved into the nuances of developing shared libraries for different contexts. When the shared library is intended for use within a single project and compiled with the same compiler, shared objects (or DLLs) with C++ interfaces can be suitable, albeit with caution around singletons and global state to avoid multithreading issues and unpredictable initialization order.

However, for wider distribution where the end user's compiler or programming language might differ, using C++ shared libraries directly is less advisable due to the instability of the C++ ABI across different compilers and versions. To overcome this, we discussed creating a C wrapper around the C++ code, leveraging the stable C ABI for broader compatibility and cross-language functionality.

We provided a comprehensive example using a `MessageSender` class, illustrating how to create a C++ library and its corresponding C wrapper. The example emphasized safe memory management by ensuring allocation and deallocation within the same binary and handling exceptions gracefully by representing them with an enumerated status in the C interface.

By following these guidelines, developers can create robust, maintainable, and widely compatible shared libraries, ensuring their usability across various platforms and programming environments. This chapter equips developers with the necessary knowledge to address common caveats and implement best practices in shared library development, fostering effective and reliable software solutions.

In the next chapter, we will shift our focus to code formatting, exploring best practices for creating clear, consistent, and readable code, which is essential for collaboration and long-term maintenance.

## 9

Code Formatting and Naming Conventions

In the vast and complex landscape of software development, some topics may seem less significant at first glance, yet they hold enormous value when considered in the broader context of creating robust and maintainable software. Code formatting is one such topic. While it might appear to be a mere aesthetic concern, it plays an essential role in enhancing code readability, simplifying maintenance, and fostering effective collaboration among team members. The significance of these aspects becomes even more pronounced in languages such as C++, where the structure and syntax can easily become complex.

In this chapter, we will delve deep into the nuances of code formatting, providing you with a comprehensive understanding of its importance. But understanding the "why" is only the first step; it's equally crucial to know the "how." Therefore, we will also explore the various tools available for automatically formatting your C++ code, taking a close look at their features and possibilities, as well as how they can be configured to meet your project's specific needs. From industry-standard tools such as Clang-Format to editor-specific plugins, we'll examine how to make these powerful utilities work for you.

By the end of this chapter, you'll have not only a thorough understanding of why code formatting is essential but also the practical knowledge to implement consistent and effective formatting across your C++ projects. So, let's turn the page and embark on this enlightening journey.

Why is code formatting important?

The importance of code formatting in software development, especially in languages such as C++, can't be overstated. Let's begin with readability, which is crucial because code is often read more frequently than it is written. Proper indentation and spacing give the code a visual structure, facilitating a quick understanding of its flow and logic. In a well-formatted code base, it's easier to scan through the code to identify key elements such as loops, conditionals, and sections. This, in turn, reduces the need for excessive comments since the code often becomes self-explanatory.

When it comes to maintainability, consistent code formatting is a boon. Well-structured code is easier to debug. For instance, a consistent indentation can quickly highlight unclosed brackets or scope issues, making it easier to spot errors. Well-formatted code also enables developers to isolate sections of code more effectively, which is essential for both debugging and refactoring. Additionally, maintainability is not just about the here and now; it's about future-proofing the code. As the code base evolves, a consistent formatting style ensures that new additions are easier to integrate.

Collaboration is another area where consistent code formatting plays a significant role. In a team setting, having a unified code style reduces the cognitive load for each team member. It allows developers to focus more on the logic and implementation of the code rather than getting sidetracked by stylistic inconsistencies. This is particularly beneficial during code reviews, where the uniform style enables reviewers to focus on the core logic and potential issues instead of being distracted by varying formatting styles. For new team members, a consistently formatted code base can be much easier to understand, helping them get up to speed more quickly.

Moreover, code formatting plays a role in quality assurance and can be automated to some extent. Many teams utilize automated formatting tools to ensure that the code base maintains a consistent style, which not only reduces the likelihood of human error but can also be a factor in code quality metrics. Automated checks for code formatting can be integrated into the CI/CD pipeline, making it a part of the overall best practices for the project.

Finally, let's not forget the impact of code formatting on version control. A consistent coding style ensures that version histories and diffs accurately reflect changes in code logic, not just style adjustments. This makes it easier to track changes, identify issues, and understand the evolution of the code base over time using tools such as `git blame` and `git history`.

In conclusion, proper code formatting serves both a functional and aesthetic purpose. It enhances readability, simplifies maintenance, and fosters collaboration, all of which contribute to the effective and efficient development of robust and maintainable software.

Overview of existing tools that facilitate compliance with coding conventions

The world of C++ development has seen an ever-increasing focus on writing clean, maintainable code. One of the cornerstones of this approach is adherence to well-defined coding conventions. Thankfully, several tools can help automate this process, making it easier for developers to focus on solving actual problems rather than fretting over code aesthetics. In this section, we'll take a broad look at some of the most popular and widely used tools for enforcing coding conventions in C++ projects.

cpplint

cpplint is a Python-based tool that aims to check your C++ code against Google's style guide, providing a less flexible but highly focused toolset for maintaining coding conventions. If you or your team admire Google's C++ coding standards, cpplint offers a straightforward path to ensure compliance within your project.

cpplint comes with a set of predefined checks based on Google's C++ style guide. These checks cover a variety of aspects, from file headers to indentation, and from variable naming to the inclusion of unnecessary headers. The tool is executed from the command line, and its output offers clear guidance on which parts of the code violate the guidelines, often providing hints on how to correct these issues.

Being Python-based, cpplint enjoys the advantage of being cross-platform. You can easily integrate it into development environments across Windows, macOS, and Linux, making it a convenient choice for diverse teams.

The command-line nature of cpplint allows it to be easily integrated into a variety of development pipelines. It can be included in pre-commit hooks, part of a CI system, or even be set to run at specific intervals during development. Several IDEs and text editors also provide plugins to run cpplint automatically on file save or during a build.

While it doesn't offer the same level of customization as some other tools, cpplint has the advantage of being backed by Google, and it follows a widely respected style guide. The tool has extensive documentation that not only explains how to use cpplint but also dives into the reasoning behind specific coding conventions, offering valuable insights into the principles of writing clean, maintainable C++ code.

The primary limitation of cpplint is its lack of flexibility. The tool is designed to enforce Google's coding standards and offers limited scope for customization. This can be a drawback if your project has unique formatting requirements or if you're working within a team that has already adopted a different set of conventions.

In conclusion, cpplint serves as a focused tool for C++ developers who wish to adopt Google's C++ style guide within their projects. While it may not offer the wide range of customization features found in some other tools, its simplicity, ease of integration, and adherence to well-respected coding standards make it a valuable asset for many development teams.

More information about cpplint can be found on the official page (https://github.com/google/styleguide/tree/gh-pages/cpplint) and in the forked GitHub repository maintained by enthusiasts (https://github.com/cpplint/cpplint).

Artistic Style

In the realm of code formatting tools, **Artistic Style (Astyle)** holds a unique position. It is designed to be a fast, small, and, above all, simple tool that supports multiple programming languages, including C++. One of the standout features of Astyle is its ease of use, making it a particularly good choice for smaller projects or for teams who are venturing into the world of automated code formatting for the first time.

Astyle provides a set of predefined styles such as ANSI, GNU, and Google, among others, which can serve as good starting points for your project's coding conventions. Additionally, it offers options to adjust indentation, align variables and pointers, and even sort modifiers, among other things. These can be controlled through command-line options or a configuration file.

A major benefit of Astyle is its cross-platform nature. It can be used on Windows, macOS, and Linux, making it a versatile choice for teams with diverse development environments.

One of the strong suits of Astyle is its ease of integration into various development pipelines. It can be easily hooked into pre-commit scripts, integrated into the most popular text editors, and even added to your CI process.

Although it may not have as extensive a community as some other tools, Astyle has been around for quite some time and has built up a solid user base. Its documentation is straightforward to understand, providing clear guidance even for those who are new to the concept of automated code formatting.

While Astyle is feature-rich, it's worth noting that it might not be the best fit for extremely large or complex projects that require highly specialized formatting rules. It offers fewer customization options compared to some other tools, which could be a limitation if your project has very specific formatting requirements.

In summary, Astyle serves as a robust and easy-to-use tool for automating code formatting in C++ projects. Its simplicity, ease of integration, and cross-platform support make it an attractive option for many developers. Whether you are new to automated code formatting or looking for a simpler alternative, Astyle offers a straightforward way to ensure that your code base adheres to consistent coding conventions. For more information, please refer to the project's official page: `https://astyle.sourceforge.net/astyle.html`.

Uncrustify

When it comes to the realm of code formatting in C++, Uncrustify stands out for its incredible range of customization options. This powerful tool offers a level of granularity that few other formatters can match, making it an ideal choice for large and complex projects with highly specific formatting needs. If you're someone who relishes the ability to fine-tune every aspect of your code's appearance, then Uncrustify is worth a closer look.

Uncrustify supports an extensive set of formatting options, allowing developers to customize everything from indentation levels and brace styles to the alignment of comments and code constructs. All these options can be set in a configuration file that can then be shared across a development team to ensure consistent formatting.

Uncrustify is cross-platform compatible and can be easily used in development environments on Windows, macOS, and Linux. It is not tied to any specific development environment and offers a variety of integration paths. It can be set up as a pre-commit hook in your version control system, integrated into popular IDEs through plugins, or even included as a step in your CI pipeline. Because of its command-line nature, integrating Uncrustify into various tools and workflows is typically straightforward.

Uncrustify has an active community and its documentation, although sometimes considered dense, is comprehensive. This provides developers with a rich source of information for understanding the tool's extensive capabilities. While the configuration can be challenging due to its sheer volume of options, numerous online resources and forums offer guidance, tips, and best practices for making the most of Uncrustify's features.

The most notable limitation of Uncrustify is its complexity. The tool's strength – its myriad customization options – can also be a weakness, especially for smaller projects or teams that don't require such a high level of configurability. Additionally, the steep learning curve could be a barrier for teams looking for a quick solution to implement consistent code formatting.

In summary, Uncrustify offers an unmatched level of customization for those looking to fine-tune their C++ code formatting to the nth degree. Its wide array of features, coupled with extensive documentation and an active community, makes it a robust choice for teams seeking to enforce very specific coding standards. If you're up for the challenge of mastering its many options, Uncrustify can serve as an invaluable tool for maintaining a clean and consistent code base. For more detailed information, please refer to the official GitHub page: `https://github.com/uncrustify/uncrustify`.

Editor plugins

In an age where development teams are more diverse than ever, relying on a single IDE for code formatting can be problematic. Not only does it force developers to adapt to a specific work environment – potentially hindering their performance – but it also creates challenges in maintaining a consistent code style across different IDEs. Furthermore, such reliance poses complications for integrating code formatting into CI/CD pipelines. This is where editor plugins come into play as a more flexible and universal solution.

One of the key advantages of editor plugins is their wide availability across multiple text editors and IDEs. Whether your team prefers Visual Studio Code, Sublime Text, Vim, or Emacs, chances are there's a plugin available that integrates with your chosen code formatting tool. This means every team member can work in the development environment they are most comfortable with, without compromising on code consistency.

ditor plugins often act as wrappers around standalone formatting tools such as Clang-Format, Astyle, and Uncrustify. This facilitates an easy transition, especially if your team is already using one of these tools. The configuration files for these tools can be shared, ensuring that the same formatting rules are applied irrespective of the editor being used.

Since many editor plugins leverage standalone command-line tools for code formatting, they naturally fit well into CI/CD pipelines. This eliminates the need to rely on IDE-specific tools, which may not be easily adaptable to CI/CD systems. With a standalone tool, the same formatting checks can be performed both locally by developers and automatically within the CI/CD pipeline, ensuring consistency across the board.

While editor plugins offer a versatile approach to code formatting, they do come with their own set of limitations. First, not all editors may support the full range of available formatting tools, although most popular editors have a wide range of plugins. Second, while installing and configuring plugins is generally straightforward, it does require an initial setup effort from each developer on the team.

Editor plugins provide an accessible and universal solution for implementing code formatting across diverse development environments. Their flexibility allows team members to choose their preferred editors without sacrificing code consistency, and their compatibility with standalone formatting tools makes them an excellent fit for inclusion in CI/CD pipelines. For teams that prioritize both developer autonomy and code consistency, editor plugins offer a balanced and effective approach.

Clang-Format

When it comes to discussing code formatting tools that have gained significant traction in the C++ community, Clang-Format undoubtedly takes a front seat. Often considered the Swiss Army knife of code formatting, this tool combines robustness with a wealth of customization options. As this chapter's favorite, we will delve deeper into its intricacies, exploring its extensive features and configurations in subsequent sections.

At its core, Clang-Format is designed to automatically reformat code so that it complies with a set of specified rules. These rules can range from handling whitespace and indentation to more complex aspects such as code block alignment and comment reformatting. Configuration is usually done via a `.clang-format` file, where developers can define their style preferences in a structured manner.

Clang-Format offers excellent cross-platform support, functioning seamlessly on Windows, macOS, and Linux. This ensures that irrespective of the development environment, your team can benefit from consistent code formatting.

Clang-Format shines due to its ease of integration. It can be invoked directly from the command line, included in scripts, or used via plugins in virtually any major text editor or IDE. This flexibility ensures that each developer can integrate it into their workflow, regardless of their choice of editor.

The command-line nature of Clang-Format also allows it to easily fit into CI/CD pipelines. With configuration files that can be stored and version-controlled alongside your code base, it ensures that the CI/CD system applies the same formatting rules as any developer would locally.

With strong backing from a broad community of developers and extensive documentation, Clang-Format has a wealth of resources available for new and experienced users alike. This community support becomes particularly beneficial when you're looking to resolve issues or customize complex formatting rules.

Given its capabilities and my personal preference for this tool, the latter sections of this chapter will dive deeper into the world of Clang-Format. From setting up your first `.clang-format` file to exploring some of its more advanced features, we'll cover how to make the most of what this powerful tool has to offer.

Clang-Format configuration – a deep dive into customizing your formatting rules

When it comes to configuring Clang-Format, the possibilities are almost endless, allowing you to tweak even the most minute details of your code's appearance. However, for those who are new to this tool or those who wish to quickly adopt a widely accepted set of rules, Clang-Format allows you to derive configurations from existing presets. These presets serve as solid foundations upon which you can build a tailored formatting style that suits your project's specific needs.

Leveraging existing presets

Clang-Format offers several built-in presets that adhere to popular coding standards. These include the following:

- `LLVM`: Adheres to the LLVM coding standards
- `Google`: Follows Google's C++ style guide
- `Chromium`: Based on Chromium's style guide, a variant of Google's style guide
- `Mozilla`: Follows the Mozilla coding standards
- `WebKit`: Adheres to the WebKit coding standards

To use one of these presets, simply set the `BasedOnStyle` option in your `.clang-format` configuration file, like so:

```
BasedOnStyle: Google
```

This tells Clang-Format to apply the Google C++ style guide as a base and then apply any additional customizations you specify.

Extending and overriding presets

After choosing a preset that aligns closest with your team's coding philosophy, you can start customizing specific rules. The `.clang-format` file allows you to override or extend the preset's rules by listing them under the `BasedOnStyle` option. For example, an extended `.clang-format` example can demonstrate how to fine-tune various aspects of code formatting. The following is a sample configuration file that starts with Google's style as a base and then customizes several specific aspects, such as indentation width, brace wrapping, and the alignment of consecutive assignments:

```
---
BasedOnStyle: Google

# Indentation
IndentWidth: 4
TabWidth: 4
UseTab: Never

# Braces
BreakBeforeBraces: Custom
BraceWrapping:
  AfterClass: true
  AfterControlStatement: false
  AfterEnum: true
  AfterFunction: true
  AfterNamespace: true
  AfterStruct: true
  AfterUnion: true
  BeforeCatch: false
  BeforeElse: false

# Alignment
AlignAfterOpenBracket: Align
AlignConsecutiveAssignments: true
AlignConsecutiveDeclarations: true
AlignOperands: true
AlignTrailingComments: true

# Spaces and empty lines
SpaceBeforeParens: ControlStatements
SpaceInEmptyParentheses: false
SpacesInCStyleCastParentheses: false
SpacesInContainerLiterals: true
SpacesInSquareBrackets: false
MaxEmptyLinesToKeep: 2
```

```
# Column limit
ColumnLimit: 80
```

Let's take a closer look at the options we chose here:

1. `IndentWidth` and `TabWidth`: These set the number of spaces for indentation and tabs, respectively. Here, `UseTab: Never` specifies not to use tabs for indentation.

2. `BreakBeforeBraces` and `BraceWrapping`: These options customize when to break before opening braces in various situations such as classes, functions, and namespaces.

3. `AlignAfterOpenBracket`, `AlignConsecutiveAssignments`, and so on: These control how various code elements, such as open brackets and consecutive assignments, are aligned.

4. `SpaceBeforeParens`, `SpaceInEmptyParentheses`, and so on: These manage spaces in different scenarios, such as before parentheses in control statements or within empty parentheses.

5. `MaxEmptyLinesToKeep`: This option limits the maximum number of consecutive empty lines to keep.

6. `ColumnLimit`: This option sets a column limit per line to ensure the code doesn't exceed the specified limit, enhancing readability.

The `.clang-format` file should be placed in the root directory of your project and committed to your version control system so that every team member and your CI/CD pipeline can use the same configuration for consistent code formatting.

Ignoring specific lines with Clang-Format

While Clang-Format is an excellent tool for maintaining a consistent coding style across a project, there might be occasions when you'd prefer to keep certain lines or blocks of code untouched. Thankfully, Clang-Format provides the functionality to exclude specific lines or code blocks from formatting. This can be particularly useful for lines where the original formatting is essential for readability or lines that contain generated code that should not be altered.

To ignore a particular line or block of code, you can use special comment markers. Place `// clang-format off` before the line or block of code that you want to ignore, and then use `// clang-format on` after the line or block to resume normal formatting. Here's an example:

```
int main() {
    // clang-format off
    int    variableNameNotFormatted=42;
    // clang-format on

    int properlyFormattedVariable = 43;
}
```

In this example, Clang-Format will not touch int   variableNameNotFormatted=42;, but will apply the specified formatting rules to int   properlyFormattedVariable = 43;.

This feature offers a fine-grained level of control over the formatting process, allowing you to combine the benefits of automated formatting with the nuances that may be required for specific coding situations. Feel free to include this in your chapter to provide a complete view of what Clang-Format offers for code style management.

Endless options for configuration

Since Clang-Format is based on the Clang compiler's code parser, it can provide the most precise analysis of source code and, as a result, the most endless configuration options. The complete list of possible settings can be found on the official page: `https://clang.llvm.org/docs/ClangFormatStyleOptions.html`.

Version control and sharing

It's generally a good practice to include your .clang-format file in your project's version control system. This ensures that every member of your team, as well as your CI/CD system, uses the same set of formatting rules, leading to a more consistent and maintainable code base.

Integrating Clang-Format into the build system

In today's software development landscape, CMake stands as a de facto industry standard for build systems. It offers a powerful and flexible way to manage builds across different platforms and compilers. Integrating Clang-Format – a tool for automatically formatting C++ code – into your CMake build process can help ensure consistent code formatting across your project. In this section, we'll delve into how this can be achieved effectively.

First, you must start by identifying the Clang-Format executable on your system using CMake's find _ program() function:

```
# Find clang-format
find _ program(CLANG _ FORMAT _ EXECUTABLE NAMES clang-format)
```

Next, you must gather all the source files you wish to format. The file(GLOB _ RECURSE ...) function is useful for this purpose:

```
# Gather all source files from the root directory recursively
file(GLOB _ RECURSE ALL _ SOURCE _ FILES
    *.cpp
    *.cc
```

```
    *.c++
    *.c
    *.C
    *.h
    *.hpp
    *.hxx
)
```

However, there's a slight hiccup: this approach also includes files from your build directory, which you likely don't want to format. This also usually applies to third-party directories. Fortunately, you can filter these out using CMake's list(FILTER ...) function:

```
# Exclude files in the build directory
list(FILTER ALL_SOURCE_FILES EXCLUDE REGEX "^${CMAKE_BINARY_DIR}.*")
```

Finally, you must create a custom CMake target that, when built, runs Clang-Format on your gathered source files:

```
# Create custom target to run clang-format
if(CLANG_FORMAT_EXECUTABLE)
    add_custom_target(
        clang-format
        COMMAND ${CLANG_FORMAT_EXECUTABLE} -i -style=file ${ALL_
SOURCE_FILES}
        COMMENT "Running clang-format"
    )
else()
    message("clang-format not found! Target 'clang-format' will not be
available.")
endif()
```

By doing this, you can create a custom target named clang-format that developers can run to automatically format all the source files in the project while ignoring any files in the build directory. Executing this target can be done with a simple make clang-format or cmake --build . --target clang-format command, ensuring that consistent formatting is maintained with ease.

Including this Clang-Format and CMake integration in your build process not only helps in maintaining a consistent coding style but also facilitates easier code reviews and collaborative development. Feel free to incorporate these insights and code snippets into your project or any technical documentation you're working on.

Clang-Format report examples

Let's prepare a simple example to demonstrate the Clang-Format tool in action. We will create a basic C++ source file named `main.cpp` with some formatting issues. Then, we will run Clang-Format on this file to automatically correct the formatting and generate a report of the changes made:

```cpp
#include <iostream>

class Sender {
public:
    void send(const std::string& message) {
        std::cout << "Sending: " << message << std::endl;
    }
};

class Receiver {
public:
    void receive(const std::string& message) {
        std::cout << "Receiving: " << message << std::endl;
    }
};

class Mediator {
public:
    Mediator(Sender sender, Receiver receiver)
        : sender_{std::move(sender)}, receiver_{std::move(receiver)} {}
    void send(const std::string& message) {
        sender_.send(message);
    }

    void receive(const std::string& message) {
        receiver_.receive(message);
    }
private:
    Sender sender_;
    Receiver receiver_;
};
```

We will try to analyze it with the Clang-Format tool and a ruleset we defined earlier in `.clang-format`:

```
make check-clang-format
[100%] Checking code format with clang-format
/home/user/clang-format/clang_format.cpp:4:2: error: code should be
```

```
clang-formatted [-Wclang-format-violations]
{
 ^

/home/user/clang-format/clang_format.cpp:6:42: error: code should be
clang-formatted [-Wclang-format-violations]
    void send(const std::string& message){
                                          ^

/home/user/clang-format/clang_format.cpp:7:18: error: code should be
clang-formatted [-Wclang-format-violations]
        std::cout<< "Sending: " <<message<< std::endl;
                 ^

/home/user/clang-format/clang_format.cpp:7:35: error: code should be
clang-formatted [-Wclang-format-violations]
        std::cout<< "Sending: " <<message<< std::endl;
                                ^

/home/user/clang-format/clang_format.cpp:7:42: error: code should be
clang-formatted [-Wclang-format-violations]
        std::cout<< "Sending: " <<message<< std::endl;
                                       ^

/home/user/clang-format/clang_format.cpp:11:6: error: code should be
clang-formatted [-Wclang-format-violations]
class
     ^

/home/user/clang-format/clang_format.cpp:12:9: error: code should be
clang-formatted [-Wclang-format-violations]
Receiver {
        ^

/home/user/clang-format/clang_format.cpp:12:11: error: code should be
clang-formatted [-Wclang-format-violations]
Receiver {
          ^

/home/user/clang-format/clang_format.cpp:14:36: error: code should be
clang-formatted [-Wclang-format-violations]
    void receive(const std::string&message){
                                       ^

/home/user/clang-format/clang_format.cpp:14:44: error: code should be
clang-formatted [-Wclang-format-violations]
    void receive(const std::string&message){
                                           ^

/home/user/clang-format/clang_format.cpp:14:45: error: code should be
clang-formatted [-Wclang-format-violations]
    void receive(const std::string&message){
                                            ^

/home/user/clang-format/clang_format.cpp:16:6: error: code should be
clang-formatted [-Wclang-format-violations]
    }};
```

```
/home/user/clang-format/clang_format.cpp:18:15: error: code should be
clang-formatted [-Wclang-format-violations]
class Mediator{
              ^

/home/user/clang-format/clang_format.cpp:18:16: error: code should be
clang-formatted [-Wclang-format-violations]
class Mediator{
               ^

/home/user/clang-format/clang_format.cpp:20:28: error: code should be
clang-formatted [-Wclang-format-violations]
    Mediator(Sender sender,Receiver receiver)
                           ^

/home/user/clang-format/clang_format.cpp:21:69: error: code should be
clang-formatted [-Wclang-format-violations]
        : sender_{std::move(sender)}, receiver_{std::move(receiver)} {}
                                                                        ^

/home/user/clang-format/clang_format.cpp:21:71: error: code should be
clang-formatted [-Wclang-format-violations]
        : sender_{std::move(sender)}, receiver_{std::move(receiver)} {}

  ^

/home/user/clang-format/clang_format.cpp:22:44: error: code should be
clang-formatted [-Wclang-format-violations]
    void send(const std::string& message) {sender_.send(message);}
                                            ^

/home/user/clang-format/clang_format.cpp:22:66: error: code should be
clang-formatted [-Wclang-format-violations]
    void send(const std::string& message) {sender_.send(message);}
                                                                  ^

/home/user/clang-format/clang_format.cpp:24:47: error: code should be
clang-formatted [-Wclang-format-violations]
    void receive(const std::string& message) {
                                              ^

/home/user/clang-format/clang_format.cpp:25:36: error: code should be
clang-formatted [-Wclang-format-violations]
        receiver_.receive(message);
                                   ^

/home/user/clang-format/clang_format.cpp:26:6: error: code should be
clang-formatted [-Wclang-format-violations]
    }
     ^

/home/user/clang-format/clang_format.cpp:28:11: error: code should be
clang-formatted [-Wclang-format-violations]
    Sender sender_;
          ^
```

```
make[3]: *** [CMakeFiles/check-clang-format.dir/build.make:71: CMakeFiles/
check-clang-format] Error 1
make[2]: *** [CMakeFiles/Makefile2:139: CMakeFiles/check-clang-format.dir/
all] Error 2
make[1]: *** [CMakeFiles/Makefile2:146: CMakeFiles/check-clang-format.dir/
rule] Error 2
make: *** [Makefile:150: check-clang-format] Error 2
```

As you can see, the errors are not very descriptive. However, most of the time, developers can understand what's wrong with the code. The tool is not only able to detect the issues but also fix them. Let's run the tool to fix the formatting issues make clang-format and see the results:

```cpp
#include <iostream>

class Sender
{
    public:
     void send(const std::string& message)
     {
         std::cout << "Sending: " << message << std::endl;
     }
};

class Receiver
{
    public:
     void receive(const std::string& message)
     {
         std::cout << "Receiving: " << message << std::endl;
     }
};

class Mediator
{
    public:
     Mediator(Sender sender, Receiver receiver)
         : sender_{std::move(sender)}, receiver_{std::move(receiver)}
     {
     }
     void send(const std::string& message) { sender_.send(message); }

     void receive(const std::string& message) { receiver_.
receive(message); }

    private:
```

```
    Sender    sender_;
    Receiver receiver_;
};
```

The code is now properly formatted and can be used in the project. This example can be included in your chapter to demonstrate the practical application of Clang-Format in a real-world scenario. In the future, developers may add more formatting rules to the `.clang-format` file and re-format the whole project by running the `make clang-format` command.

Extending for code format checks for CI

When setting up CI pipelines, it's often beneficial to only check whether the code complies with the established formatting rules rather than automatically modifying the source files. This ensures that any code that doesn't meet the style guidelines is flagged, prompting the developer to fix it manually. Clang-Format supports this use case with the `--dry-run` and `--Werror` options, which, when combined, cause the tool to exit with a non-zero status code if any file is reformatted.

You can extend the existing CMake setup so that it includes a new custom target that only checks the code format. Here's how to do this:

```
# Create custom target to check clang-format
if(CLANG_FORMAT_EXECUTABLE)
    add_custom_target(
        check-clang-format
        COMMAND ${CLANG_FORMAT_EXECUTABLE} -style=file -Werror --dry-run ${ALL_SOURCE_FILES}
        COMMENT "Checking code format with clang-format"
    )
else()
    message("clang-format not found! Target 'check-clang-format' will not be available.")
endif()
```

In this extended setup, a new custom target named `check-clang-format` has been added. The `--dry-run` option ensures that no files are modified, while `-Werror` causes Clang-Format to exit with an error code if any formatting discrepancies are found. This target can be run with `make check-clang-format` or `cmake --build . --target check-clang-format`.

Now, in your CI pipeline script, you can invoke this custom target to enforce code-style checks. If the code is not formatted according to the guidelines specified, the build will fail, alerting the team that there is a formatting issue that needs to be resolved.

For example, in our `.clang-format` file, we set the indent width to four spaces, but the `main.cpp` file only uses two:

```
int main() {
  return 0;
}
```

Once we run the checker, it shows the problematic code without changing it:

```
make check-clang-format

-- Configuring done
-- Generating done
-- Build files have been written to: /home/user/clang-format-tidy/build
[100%] Checking code format with clang-format
/home/user/clang-format-tidy/main.cpp:2:13: error: code should be clang-
formatted [-Wclang-format-violations]
int main() {
            ^
make[3]: *** [CMakeFiles/check-clang-format.dir/build.make:71: CMakeFiles/
check-clang-format] Error 1
make[2]: *** [CMakeFiles/Makefile2:137: CMakeFiles/check-clang-format.dir/
all] Error 2
make[1]: *** [CMakeFiles/Makefile2:144: CMakeFiles/check-clang-format.dir/
rule] Error 2
make: *** [Makefile:150: check-clang-format] Error 2
```

By adding this custom target to your CMake setup, you add an additional layer of quality assurance to your project. It ensures that no code that violates the established formatting guidelines can make its way into the code base unnoticed. This is particularly helpful in collaborative environments where multiple developers might be contributing to the same project. Feel free to include this advanced example and its rationale in your technical content.

Clang-Format support across various editors

Clang-Format enjoys extensive support across a multitude of text editors and IDEs, streamlining the code-formatting process irrespective of your development environment. One of the significant advantages of integrating Clang-Format directly into your IDE or text editor is the ability to invoke it effortlessly, right from your development environment. Even better, many editors support automatically triggering Clang-Format upon saving a file. This feature can be a massive boon to productivity and code quality as it ensures that every saved version of a source file adheres to the project's coding standards without requiring manual intervention.

In Visual Studio Code, there are a few plugins that provide integration with Clang-Format:

- **C/C+**, by Microsoft: `https://marketplace.visualstudio.com/items?itemName=ms-vscode.cpptools`

- **Clang-Format**, by Xaver Hellauer: `https://marketplace.visualstudio.com/items?itemName=xaver.clang-format`

- **ClangD**, by the LLVM (the creators of Clang, Clang-Format, and other tools): `https://marketplace.visualstudio.com/items?itemName=llvm-vs-code-extensions.vscode-clangd`

Vim and NeoVim users can leverage plugins such as `vim-clang-format` to integrate Clang-Format, and even map it to specific keyboard shortcuts for quick formatting. Also, it can usually be enabled via an LSP provider plugin or feature.

For developers using the full-fledged version of Visual Studio, Clang-Format integration is built-in; you can easily specify a `.clang-format` configuration file and the IDE will use it when you format your code.

Similarly, JetBrains' CLion supports Clang-Format out of the box, allowing users to import `.clang-format` configuration files directly into the project settings. This broad range of editor support makes it effortless to maintain consistent code formatting across diverse development teams as each team member can use their preferred tools without compromising on code quality.

Checking name styling

After meticulously formatting our code to ensure that spaces, asterisks, alignments, and brace positions are all in place, there remains one final frontier to unify – naming style. Ensuring consistency in naming conventions across classes, variables, functions, and other identifiers can often be a painstaking process, usually relegated to vigilant peer reviews. However, there's an automated way to achieve this, thereby reducing manual effort and error.

Clang-Tidy comes to the rescue for this purpose. While we'll dive deeper into the various capabilities of Clang-Tidy in the next chapter, it's worth noting that it's more than just a linter. It offers a plethora of checks for not just syntactic sugar but also semantic analysis and readability. One of its most useful features in the context of naming conventions is the identifier naming check. By configuring this check, you can enforce project-wide rules for naming various entities.

Let's say you want your class, struct, and enum names to be `CamelCase`, your namespaces, variables, functions, and methods to be `lower _ case`, and your constants to be `UPPER _ CASE`. Additionally, you prefer that private and protected variables have a trailing underscore, `_` , while public ones do not. All of these requirements can be configured in a simple `.clang-tidy` file, which Clang-Tidy will read to enforce your naming rules:

```
---
Checks:          'readability-identifier-naming'
```

```
FormatStyle:      file

CheckOptions:
  - key: readability-identifier-naming.NamespaceCase
    value: 'lower _ case'
  - key: readability-identifier-naming.InlineNamespaceCase
    value: 'lower _ case'
  - key: readability-identifier-naming.EnumCase
    value: 'CamelCase'
  - key: readability-identifier-naming.EnumConstantCase
    value: 'UPPER _ CASE'
  - key: readability-identifier-naming.ClassCase
    value: 'CamelCase'
  - key: readability-identifier-naming.StructCase
    value: 'CamelCase'
  - key: readability-identifier-naming.ClassMethodCase
    value: 'lower _ case'
  - key: readability-identifier-naming.FunctionCase
    value: 'lower _ case'
  - key: readability-identifier-naming.VariableCase
    value: 'lower _ case'
  - key: readability-identifier-naming.GlobalVariableCase
    value: 'lower _ case'
  - key: readability-identifier-naming.StaticConstantCase
    value: 'UPPER _ CASE'
  - key: readability-identifier-naming.PublicMemberCase
    value: 'lower _ case'
  - key: readability-identifier-naming.ProtectedMemberCase
    value: 'lower _ case'
  - key: readability-identifier-naming.PrivateMemberCase
    value: 'lower _ case'
  - key: readability-identifier-naming.PrivateMemberSuffix
    value: ' _ '
  - key: readability-identifier-naming.ClassMemberCase
    value: 'lower _ case'
```

These rules can be endlessly extended with the highest resolution. The full documentation for existing checks is available at https://clang.llvm.org/extra/clang-tidy/checks/readability/identifier-naming.html.

By incorporating Clang-Tidy into your build process and CI pipeline, you can automate the enforcement of these naming conventions, making the code base easier to read, maintain, and collaborate on. We will delve deeper into configuring and using Clang-Tidy for various other checks in the upcoming chapter.

Integrating Clang-Tidy into the build system

We can adapt our existing CMake setup so that it includes Clang-Tidy checks, similar to what we did with Clang-Format. Here's a sample CMake script that sets up custom targets for running Clang-Tidy on a C++ project:

```
# Generate compilation database in the build directory
set(CMAKE_EXPORT_COMPILE_COMMANDS ON)

# Find clang-tidy
find_program(CLANG_TIDY_EXECUTABLE NAMES clang-tidy)

# Gather all source files from the root directory recursively
file(GLOB_RECURSE ALL_SOURCE_FILES
    *.cpp
    *.cc
    *.c++
    *.c
    *.C
    *.h
    *.hpp
    *.hxx
)

# Exclude files in the build directory
list(FILTER ALL_SOURCE_FILES EXCLUDE REGEX "^${CMAKE_BINARY_DIR}.*")

# Create custom target to run clang-tidy
if(CLANG_TIDY_EXECUTABLE)
    add_custom_target(
        clang-tidy
        COMMAND ${CLANG_TIDY_EXECUTABLE} -p=${CMAKE_BINARY_DIR}
${ALL_SOURCE_FILES}
        COMMENT "Running clang-tidy"
    )
else()
    message("clang-tidy not found! Target 'clang-tidy' will not be
available.")
endif()

# Create custom target to check clang-tidy
if(CLANG_TIDY_EXECUTABLE)
    add_custom_target(
        check-clang-tidy
```

```
        COMMAND ${CLANG_TIDY_EXECUTABLE} -p=${CMAKE_BINARY_DIR}
--warnings-as-errors=* ${ALL_SOURCE_FILES}
        COMMENT "Checking code quality with clang-tidy"
    )
else()
    message("clang-tidy not found! Target 'check-clang-tidy' will not be
available.")
endif()
```

In this script, we locate the `clang-tidy` executable using `find_program`. Similar to the Clang-Format setup, we then gather all the source files from the root directory recursively, making sure we exclude those in the build directory.

Two custom targets are added here:

- `clang-tidy`: This target runs Clang-Tidy on all gathered source files. The `-p=${CMAKE_BINARY_DIR}` flag specifies the build directory containing the `compile_commands.json` file, which Clang-Tidy uses for its checks. This JSON file is generated by CMake and contains information about how each source file in the project is compiled. It includes details such as the compiler options, include directories, defines, and so on. Clang-Tidy uses this information to understand the build context of each source file, allowing it to perform more accurate and meaningful checks.

- `check-clang-tidy`: This target performs the same operation but with the `--warnings-as-errors=*` flag. This will treat all warnings as errors, which is especially useful for CI/CD pipelines to ensure code quality.

As in your previous setup, running these custom targets can be done via `make clang-tidy` or `make check-clang-tidy` or their equivalent `cmake --build . --target clang-tidy` and `cmake --build . --target check-clang-tidy`.

By integrating Clang-Tidy into your CMake build process, you'll be providing another layer of automated code quality checks, much like you did with Clang-Format. Feel free to include this in your chapter for a comprehensive look at automated code quality assurance.

Checking source code name styling with Clang-Tidy

Now that we've successfully configured Clang-Tidy's rules and integrated the tool into our CMake build system, it's time for a real-world test. For this purpose, we'll use a snippet of C++ code that deliberately violates the naming conventions we've established:

```
#include <string>
#include <vector>

namespace Filesystem { // CamelCase instead of lower_case
```

```cpp
enum class Permissions : uint8_t { READ, WRITE, execute };

struct User {
    std::string name_; // redundant suffix _ for public member
    int Id = 0;         // CamelCase instead of lower_case
    Permissions permissions;
};

class file { // lower_case instead of CamelCase
public:
    file(int id, const std::string &file_name,
        const std::vector<User> access_list)
        : id{id}, FileName_{file_name}, access_list_{access_list} {}

    int GetId() const // CamelCase instead of lower_case
    {
        return id;
    }
    auto &getName() const // camelBack instead of lower_case
    {
        return FileName_;
    }

    const std::vector<User> &access_list() const { return access_list_; }

private:
    int id;                  // missing suffix _
    std::string FileName_; // CamelCase instead of lower_case
    std::vector<User> access_list_;
};

} // namespace Filesystem

int main() {

    auto user        = Filesystem::User{};
    user.name_       = "user";
    user.permissions = Filesystem::Permissions::execute;

    auto file = Filesystem::file{0, "~/home/user/file", {user}};

    return 0;
}
```

When we run `make clang-tidy`, Clang-Tidy will spring into action, scanning the offending code and flagging any naming issues directly in the terminal output. I've only provided a partial output here to save space:

```
make check-clang-tidy
[100%] Checking code quality with clang-tidy
9 warnings generated.
/home/user/clang-format-tidy/main.cpp:4:11: error: invalid case style
for namespace 'Filesystem' [readability-identifier-naming,-warnings-as-
errors]
    4 | namespace Filesystem { // CamelCase instead of lower _ case
      |           ^~~~~~~~~~
      |           filesystem
/home/user/clang-format-tidy/main.cpp:6:49: error: invalid case style for
enum constant 'execute' [readability-identifier-naming,-warnings-as-
errors]
    6 | enum class Permissions : uint8 _ t { READ, WRITE, execute };
      |                                                   ^~~~~~~
      |                                                   EXECUTE
/home/user/clang-format-tidy/main.cpp:9:17: error: invalid case style
for public member 'name _ ' [readability-identifier-naming,-warnings-as-
errors]
    9 |     std::string name _ ; // redundant suffix _ for public member
      |                 ^~~~~
      |                 name
/home/user/clang-format-tidy/main.cpp:10:9: error: invalid case style for
public member 'Id' [readability-identifier-naming,-warnings-as-errors]
   10 |     int Id = 0;        // CamelCase instead of lower _ case
      |         ^~
      |         id
/home/user/clang-format-tidy/main.cpp:14:7: error: invalid case style for
class 'file' [readability-identifier-naming,-warnings-as-errors]
   14 | class file { // lower _ case instead of CamelCase
      |       ^~~~
      |       File
   15 | public:
   16 |     file(int id, const std::string &file _ name,
      |     ~~~~
      |     File
/home/user/clang-format-tidy/main.cpp:20:9: error: invalid case style for
function 'GetId' [readability-identifier-naming,-warnings-as-errors]
   20 |     int GetId() const // CamelCase instead of lower _ case
      |         ^~~~~
      |         get _ id
9 warnings treated as errors
```

This exercise demonstrates the tangible benefits of integrating Clang-Tidy into the build process. It not only identifies deviations from established naming conventions in the code but also provides an opportunity for immediate rectification. It's an invaluable step toward maintaining a code base that is not just functional but also consistently well-structured. It's advisable to include the make clang-tidy command in your CI pipeline. By doing so, you can automatically validate the naming conventions and other code styling rules for every commit made to your repository. This will help ensure that any new contributions to the code base conform to the established guidelines. If a commit fails the Clang-Tidy checks, the CI pipeline can flag it for review, making it easier to maintain a consistent, high-quality code base. This added layer of automation eliminates the need for manual checks for these issues, thus streamlining the code review process and making your development workflow more efficient.

Fixing naming issues automatically

The real power of Clang-Tidy lies in its ability to not just identify issues but also to rectify them automatically. Manual fixes can be time-consuming and error-prone, making automation incredibly valuable in a fast-paced development environment. Fortunately, Clang-Tidy excels in this area. Most of the fixes suggested by the tool can be applied automatically, saving you countless hours of manual labor and potential errors. To apply these automatic fixes, simply run make clang-tidy in your terminal. The tool will scan the code for violations, and where possible, it will automatically correct the code so that it aligns with your configured guidelines:

```cpp
#include <string>
#include <vector>

namespace filesystem { // CamelCase instead of lower_case

enum class Permissions : uint8_t { READ, WRITE, EXECUTE };

struct User {
    std::string name_; // redundant suffix _ for public member
    int id = 0;        // CamelCase instead of lower_case
    Permissions permissions;
};

class File { // lower_case instead of CamelCase
public:
    File(int id, const std::string &file_name,
        const std::vector<User> access_list)
        : id_{id}, file_name_{file_name}, access_list_{access_list} {}

    int get_id() const // CamelCase instead of lower_case
```

```
    {
        return id _ ;
    }
    auto &get _ name() const // camelBack instead of lower _ case
    {
        return file _ name _ ;
    }

    const std::vector<User> &access _ list() const { return access _
list _ ; }

private:
    int id _ ;                    // missing suffix _
    std::string file _ name _ ;   // CamelCase instead of lower _ case
    std::vector<User> access _ list _ ;
};

} // namespace filesystem

int main() {

    auto user        = filesystem::User{};
    user.name        = "user";
    user.permissions = filesystem::Permissions::EXECUTE;

    auto file = filesystem::File{0, "~/home/user/file", {user}};

    return 0;
}
```

Note that not only were the classes, methods, and variables definitions updated but also references to them. This functionality makes Clang-Tidy not just a diagnostic tool but a valuable assistant in maintaining the overall quality of your code base.

Important caveats

There are some important caveats to consider when using Clang-Tidy. Let's take a look:

- **Single versus multiple instances**: The CMake configuration we've discussed runs a single instance of Clang-Tidy to check and fix all the source files. While this may be sufficient for smaller projects, it can become a bottleneck for larger code bases with numerous checks. In such scenarios, it might be more efficient to divide the source code into logical groups and run multiple instances of Clang-Tidy in parallel. This strategy can significantly reduce the time it takes to scan the entire code base.

- **Commit before fixing**: While Clang-Tidy's ability to automatically fix issues is invaluable, it's advised to use this feature only on code that has been committed to your version control system. Some of the checks provided by Clang-Tidy can be unstable and, in rare instances, may even introduce bugs. Committing your code beforehand ensures that you have a stable point to revert to if things go awry.

- **Syntax limitations**: Lastly, it's worth noting that Clang-Tidy may not always be up-to-date with the latest C++ syntax. While it generally does an excellent job, there could be instances where it flags false positives or is unable to handle new language features. For example, let's say the User struct was initialized using C++20 designated initializer lists, as shown in the following example:

```
auto user = Filesystem::User{
    .name _ = "user", .permissions =
Filesystem::Permissions::execute};
```

Clang-Tidy will fix the name _ variable and the execute constant in their definition but will completely ignore the initializer, which will eventually lead to a compilation error.

Being aware of these caveats allows you to employ Clang-Tidy more effectively and leverage its strengths while mitigating potential risks.

Example project

For those who wish to delve deeper into the details and get hands-on experience with the configuration and usage of Clang-Tidy and Clang-Format, an example project complete with the CMake setup and code snippets is available on GitHub (https://github.com/f-squirrel/clang-format-tidy). This will allow you to better understand the nuances and practical applications of integrating these tools into your C++ projects. Feel free to clone the repository, experiment with the code, and even contribute to enhancing it further.

Clang-Tidy support across various editors

The IDE and editor support for Clang-Tidy is broadly similar to that for Clang-Format, making it equally accessible and easy to integrate into your development workflow. The principal advantage of this integrated support is the immediate feedback loop it offers. As you code, Clang-Tidy warnings and errors will appear directly within your IDE, enabling you to spot potential issues without having to leave the development environment. This is immensely valuable for maintaining code quality in real time, rather than as a separate step.

Moreover, many IDEs also provide an interface to apply Clang-Tidy's automatic fixes directly from within the editor, making it easier than ever to adhere to your coding standards. For example, the following screenshot of Visual Studio Code illustrates the inline warnings:

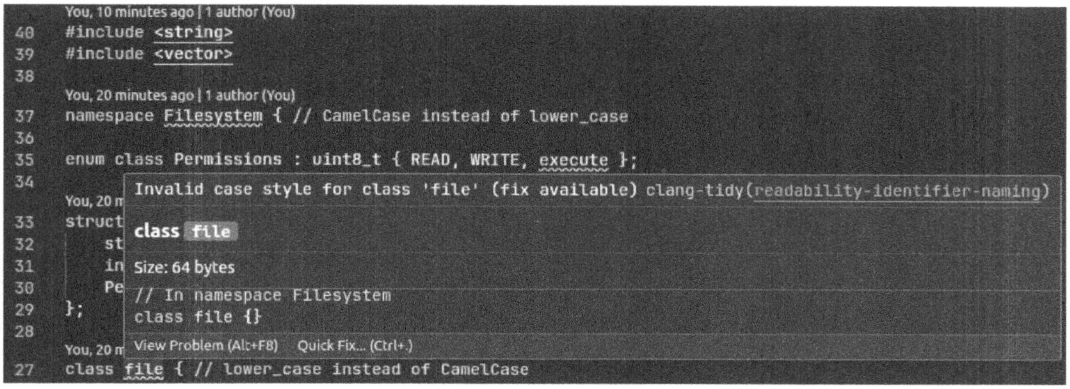

Figure 9.1 – Warnings

The following screenshot shows the fixes that can be applied to them:

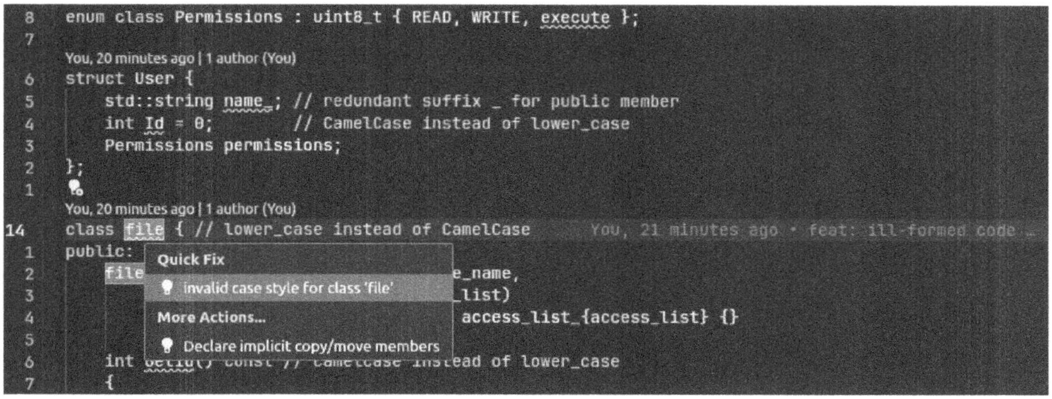

Figure 9.2 – Applicable fixes

This real-time, in-editor feedback mechanism can significantly boost your productivity and code quality, making Clang-Tidy not just a static code analysis tool but an integral part of your coding process.

Summary

In this chapter, we navigated the essential landscape of automated code quality maintenance, focusing particularly on code formatting and naming conventions. We started by providing an overview of existing tools that can help enforce coding standards, then zeroed in on Clang-Format and Clang-Tidy as comprehensive solutions for these issues. We learned not only how to use these tools to automatically check and fix our code but also how to seamlessly integrate them into our build system, CI pipelines, and code editors.

By doing so, we've laid a strong foundation for ensuring that our code remains consistent and adheres to best practices, all with minimal manual intervention. This sets the stage perfectly for the next chapter, where we will dive deeper into the realm of static code analysis, further solidifying our commitment to high-quality, maintainable code.

10

Introduction to Static Analysis in C++

In the complex and demanding world of software development, ensuring the quality and reliability of code is not just a necessity but a discipline in itself. As C++ developers, we constantly seek methodologies and tools that can aid us in this quest. This chapter is dedicated to one such powerful approach: static analysis. Renowned for being both the fastest and the cheapest way to identify bugs, static analysis stands as a pillar in the software quality assurance process. We will delve into its intricacies, explore popular tools such as Clang-Tidy, PVS-Studio, and SonarQube, and understand how to effectively integrate static analysis into your C++ development workflow.

The essence of static analysis

Static analysis is the examination of source code without executing it. This process, typically automated by various tools, involves scanning the code to identify potential errors, code smells, security vulnerabilities, and other issues. It's akin to a thorough proofreading session where the code is scrutinized for quality and reliability before it ever runs.

Why static analysis? Here are the reasons:

- **Speed and cost-effectiveness**: The foremost advantage of static analysis is its speed and cost-effectiveness. It is arguably the fastest and cheapest method to find bugs. Automating the detection of issues drastically reduces the time and effort required compared to manual code reviews and other testing methods. Catching and resolving issues early in the development cycle significantly lowers the cost of fixes, which escalates if bugs are found later in production.

- **Pre-execution bug detection**: Static analysis occurs before the code is executed, making it a proactive measure in software quality assurance. This pre-execution analysis allows developers to identify and rectify issues without the overhead of setting up testing environments or dealing with the complexities of running the code.

- **Coding standard enforcement**: It helps in maintaining a consistent coding standard, ensuring that the code base adheres to the best practices and conventions of C++ programming. This enforcement not only improves code quality but also enhances maintainability and readability.

- **Comprehensive coverage**: With the ability to scan the entire code base, static analysis provides a level of thoroughness that is challenging to achieve through manual methods. This comprehensive coverage ensures that no part of the code is overlooked.

- **Security and reliability**: Early detection of security vulnerabilities is another critical benefit. Static analysis contributes significantly to the security and reliability of the application by catching vulnerabilities that might otherwise go unnoticed until exploitation.

- **Educational aspect**: It also serves an educational purpose, enhancing developers' understanding of C++ and familiarizing them with common pitfalls and best practices.

In the subsequent sections, we'll explore how to leverage static analysis to its fullest potential in C++ projects. Following this, in the next chapter, we will compare and contrast these insights with dynamic analysis, offering a complete picture of the analysis landscape in C++ software development.

Leveraging newer compiler versions for enhanced static analysis

While the production environment often mandates specific, sometimes older, compiler versions for various reasons, including stability and compatibility, there is immense value in periodically building your project with newer versions of compilers. This practice serves as a forward-looking static analysis strategy, harnessing the advancements and improvements made in the latest compiler releases.

Newer compiler versions are frequently equipped with enhanced analysis capabilities, more sophisticated warning mechanisms, and updated interpretations of the C++ standard. They can identify issues and potential code improvements that older compilers might overlook. By compiling with these cutting-edge tools, developers can proactively discover and address latent issues in their code base, ensuring that the code remains robust and compliant with evolving C++ standards.

Additionally, this approach offers a preview of potential issues that might arise when an eventual update to the production compiler is undertaken. It provides an opportunity to future-proof the code base, making the transition to newer compiler versions smoother and more predictable.

In essence, incorporating newer compiler versions into the build process, even if they are not used for production builds, is a strategic measure. It not only elevates the quality of the code through advanced static analysis but also prepares the code base for future technological shifts, ensuring a state of continuous improvement and readiness for advancement.

Compiler settings to harden C++ code

In the pursuit of robust and secure C++ code, configuring compiler settings plays a pivotal role. Compiler flags and options can significantly enhance code quality by enabling stricter error checking, warnings, and security features. This section focuses on recommended settings for three major compilers in the C++ ecosystem: the **GNU Compiler Collection (GCC)**, Clang, and **Microsoft Visual C++ (MSVC)**. These settings are particularly valuable in a static analysis context as they enable the detection of potential issues at compile time.

GCC

GCC is known for its extensive set of options that can help harden C++ code. Key flags include the following:

- `-Wall -Wextra`: Enables most warning messages, catching potential issues such as uninitialized variables, unused parameters, and more
- `-Werror`: Treats all warnings as errors, forcing them to be addressed
- `-Wshadow`: Warns whenever a local variable shadows another variable, which can lead to confusing bugs
- `-Wnon-virtual-dtor`: Warns if a class with virtual functions has a non-virtual destructor, which can lead to undefined behavior
- `-pedantic`: Enforces strict ISO C++ compliance, rejecting non-standard code
- `-Wconversion`: Warns on implicit conversions that may alter a value, useful for preventing data loss
- `-Wsign-conversion`: Warns on implicit conversions that change the sign of a value

Clang

Clang, part of the LLVM project, shares many flags with GCC but also provides additional checks and a reputation for generating more human-readable warnings:

- `-Weverything`: Enables every warning available in Clang, providing a comprehensive check of the code. This can be overwhelming, so it's often used with selective disabling of less critical warnings.
- `-Werror, -Wall, -Wextra, -Wshadow, -Wnon-virtual-dtor,     -pedantic, -Wconversion, and -Wsign-conversion`: Similar to GCC, these flags are also applicable in Clang and serve the same purposes.
- `-Wdocumentation`: Warns about documentation inconsistencies, which is useful when maintaining large code bases with extensive comments.
- `-fsanitize=address, -fsanitize=undefined`: Enables `AddressSanitizer` and `UndefinedBehaviorSanitizer` to catch memory corruption and undefined behavior issues.

MSVC

MSVC, while having a different set of flags, also offers robust options for enhancing code safety:

- `/W4`: Enables a higher warning level, similar to `-Wall` in GCC/Clang. This includes most of the useful warnings for common issues.

- `/WX`: Treats all compiler warnings as errors.

- `/sdl`: Enables additional security checks, such as buffer overflow detection and integer overflow checks.

- `/GS`: Provides buffer security checks, helping prevent common security vulnerabilities.

- `/analyze`: Enables static code analysis to detect issues such as memory leaks, uninitialized variables, and other potential errors at compile time.

By utilizing these compiler settings, developers can significantly harden their C++ code, making it more secure, robust, and compliant with best practices. While the default settings of compilers catch many issues, enabling these additional flags ensures a much stricter and more thorough analysis of the code. It is important to note that while these settings can greatly enhance code quality, they should be complemented with good programming practices and regular code reviews for the best results. In the next chapter, we will shift our focus to dynamic analysis, another key component in ensuring the overall quality and security of C++ applications.

Static analysis via multiple compilers

In the realm of C++ development, leveraging the capabilities of compilers for static analysis is an often underutilized strategy. Compilers such as GCC and Clang come equipped with a plethora of compilation flags that enable rigorous static analysis, helping to identify potential issues without the need for additional tools. Employing these flags is not only convenient but also highly effective in enhancing code quality.

One best practice that I advocate for is building C++ projects with multiple compilers. Each compiler has its unique set of diagnostics, and by utilizing more than one, projects can gain a more comprehensive insight into potential issues. GCC and Clang are particularly notable for their similarity in supported flags, as well as their wide-ranging support for various architectures and operating systems. This compatibility makes it feasible to integrate both into a project's build process for cross-checking code.

However, incorporating these practices in a Windows environment can present additional challenges. While GCC and Clang are versatile, projects often also benefit from the distinct diagnostics provided by MSVC. MSVC integrates seamlessly with the Windows ecosystem and brings to the table a different perspective on code analysis, which can be especially beneficial for projects targeting Windows platforms. Although managing multiple compilers might introduce some complexity, the payoff in identifying a broader spectrum of potential issues is invaluable. By embracing this multi-compiler approach, projects can significantly enhance their static analysis rigor, leading to more robust and reliable C++ code.

Highlighting compiler differences – unused private members in GCC versus Clang

A nuanced understanding of the diagnostic capabilities of different compilers can be crucial in C++ development. This is exemplified in the way GCC and Clang handle unused private member variables. Consider the following class definition:

```
#include <iostream>

class NumberWrapper {
    int number;
public:
    NumberWrapper() {
    }
};
```

Here, the `number` private member in the `NumberWrapper` class is initialized but never used. This situation presents a potential issue in the code that could indicate redundancy.

Let's compare how GCC and Clang handle unused private members:

- **GCC's diagnostic approach**: In version 13, GCC does not typically warn about the `number` unused private member. This lack of warning might lead to unintentional neglect of inefficiencies or redundancies in the class design.

- **Clang's diagnostic approach**: Conversely, Clang version 17 actively flags this issue with a specific warning: `warning: private field 'number' is not used`. This precise diagnostic helps in promptly identifying and addressing potential oversights in the class's implementation.

Highlighting compiler differences – compiler checks for uninitialized variables

When dealing with class variables in C++, ensuring proper initialization is crucial to prevent undefined behavior. This aspect is highlighted in how different compilers detect uninitialized but used variables. Consider the example of the `NumberWrapper` class:

```
#include <iostream>

class NumberWrapper {
    int number;
public:
    NumberWrapper(int n) {
        (void)n; // to avoid warning: unused parameter 'n'
        std::cout << "init with: " << number << std::endl;
```

```
    }
};

int main() {
    auto num = NumberWrapper{1};
    (void) num;
    return 0;
}
```

In this code, the `number` member variable is not initialized, leading to undefined behavior when it's used in the constructor. It can print something such as `init with: 32767`.

We'll now compare the approaches used by GCC and Clang in this regard:

- **GCC's diagnostic approach**: GCC version 13 effectively flags this critical issue with a warning: `warning: 'num.NumberWrapper::number' is used uninitialized`. This warning serves as an important alert to developers, drawing attention to the risk of using uninitialized variables, which can lead to unpredictable program behavior or subtle bugs.

- **Clang's diagnostic approach**: Interestingly, Clang version 17 does not generate a warning for the same code, potentially allowing this oversight to go unnoticed in environments where only Clang is used. This demonstrates a case where relying solely on Clang might miss certain classes of errors that GCC can catch.

The two examples discussed previously offer compelling insights into the distinctive strengths and nuances of GCC and Clang's diagnostic capabilities. These instances – one highlighting Clang's ability to flag unused private fields and the other showcasing GCC's proficiency in warning about uninitialized class variables – exemplify the importance of a multi-compiler strategy in C++ development.

By utilizing both Clang and GCC, developers can harness a more comprehensive and diversified static analysis process. Each compiler, with its unique set of warnings and checks, can reveal different potential issues or optimizations. Clang, known for its detailed and specific warnings, such as flagging unused private fields, complements GCC's vigilant checks for fundamental yet critical issues such as uninitialized variables. This synergy between the compilers ensures a more thorough vetting of the code, leading to higher quality and more reliable and maintainable software.

In essence, the combination of Clang and GCC does not just add value in terms of the sum of their individual capabilities; it creates a more robust and holistic environment for static analysis. As the C++ language and its compilers continue to evolve, staying adaptable and open to multiple tools for static analysis remains a best practice for developers aiming for excellence in their craft. This approach aligns well with the ever-present goal in software development: writing clean, efficient, and error-free code.

Exploring static analysis with Clang-Tidy

As we delve deeper into the realm of static analysis, a tool that stands out for its versatility and depth is Clang-Tidy. Developed by the LLVM Foundation, the same organization behind the Clang compiler, Clang-Tidy is a linter and static analysis tool designed for C++ code. It extends beyond the capabilities of what a traditional compiler would check, offering a range of diagnostics that include stylistic errors, programming mistakes, and even subtle bugs that are often missed during regular code reviews. Previously, we explored how Clang-Tidy can be adept at code formatting; now, we will explore its prowess in static analysis, uncovering its ability to scrutinize C++ code at a level that ensures not just conformity but also excellence in coding standards.

Clang-Tidy works by using the Clang frontend to parse C++ code, enabling it to understand the code's structure and syntax in depth. This deep understanding allows Clang-Tidy to perform complex checks that go beyond mere textual analysis, examining the code's semantics and even the flow of execution. It's not just about finding syntactic discrepancies; it's about understanding the code's behavior and intent.

Categories of checks in Clang-Tidy

Clang-Tidy categorizes its checks into several groups, each targeting specific types of issues. Let's break down these categories and explore examples for each:

- **Performance checks**: Focus on identifying inefficient patterns in the code that can slow down execution; for example, unnecessary copying of objects. Clang-Tidy can flag cases where an object is copied but could be moved or passed by reference to avoid the overhead of copying:

```cpp
#include <vector>
std::vector<int> createLargeVector();
void processVector(std::vector<int> vec);

int main() {
    std::vector<int> vec = createLargeVector();
    processVector(vec); // Clang-Tidy: Use std::move to avoid copying
    return 0;
}
```

- **Modernize checks**: Aim to upgrade code to modern C++ standards, such as C++11 and beyond; for example, replacing traditional `for` loops with range-based `for` loops for better readability and safety:

```cpp
std::vector<int> myVec = {1, 2, 3};
for (std::size_t i = 0; i < myVec.size(); ++i) {
    // Clang-Tidy: Use a range-based for loop instead
    std::cout << myVec[i] << std::endl;
}
```

- **Bug detection**: Identify potential errors or logical bugs in the code; for example, detecting null pointer dereferences:

```
int* ptr = nullptr;
int value = *ptr; // Clang-Tidy: Dereference of null pointer
```

- **Style checks**: Enforce specific coding styles for consistency and readability; for example, enforcing variable naming conventions:

```
int MyVariable = 42; // Clang-Tidy: Variable name should be
lower_case
```

- **Readability checks**: Focus on making the code more understandable and maintainable; for example, simplifying complex Boolean expressions:

```
bool a, b, c;
if (a && (b || c)) {
    // Clang-Tidy: Simplify logical expression
}
```

- **Security checks**: Target potential security vulnerabilities; for example, highlighting uses of dangerous functions known to pose security risks:

```
strcpy(dest, src); // Clang-Tidy: Use of function 'strcpy' is
insecure
```

Expanding Clang-Tidy's capabilities with custom checks

Clang-Tidy's versatility is further enhanced by its support for custom checks, allowing companies and projects to tailor static analysis to their specific needs and coding standards. This customization capability has led to the creation of various categories of checks, each aligning with the guidelines of different organizations or projects. Next, we explore some notable examples:

- **Google checks (google-*)**: These checks enforce the guidelines specified in Google's C++ style guide. They include rules about naming conventions, runtime references, and build issues. For instance, `google-runtime-references` enforces Google's preference for pointers over non-const references.

- **Google's Abseil checks**: Abseil is an open source collection of C++ library code developed by Google. Checks specific to Abseil ensure adherence to the library's best practices, such as avoiding certain deprecated functions or classes.

- **Fuchsia checks**: Tailored for the Fuchsia operating system, these checks enforce coding standards and best practices specific to the Fuchsia project. They help maintain consistency and quality in the code base contributing to this OS.

- **Zircon checks**: Zircon is the core platform that powers the Fuchsia OS. Clang-Tidy includes checks tailored to Zircon's development, focusing on its unique architecture and development standards.

- **Darwin checks**: These checks are specifically designed for Darwin, the open source Unix-like operating system released by Apple. They ensure compliance with Darwin's development practices.

- **LLVM checks (llvm-*)**: These checks are designed to enforce LLVM coding standards. They are particularly useful for developers contributing to LLVM or its subprojects.

- **C++ Core Guidelines checks**: Clang-Tidy includes checks that enforce the C++ Core Guidelines, a set of best practices for writing modern C++. This includes rules for type safety, resource management, and performance.

- **Type safety checks (cppcoreguidelines-*)**: These checks ensure type safety in various scenarios. For instance, `cppcoreguidelines-pro-type-member-init` ensures that class members are properly initialized. `cppcoreguidelines-pro-type-reinterpret-cast` warns against the use of `reinterpret_cast`, encouraging safer casting alternatives.

- **Interface guidelines**: These checks enforce rules for designing clean and manageable interfaces. For example, `cppcoreguidelines-non-private-member-variables-in-classes` discourages the use of non-private member variables to maintain encapsulation.

- **Concurrency guidelines**: Checks such as `cppcoreguidelines-avoid-magic-numbers` help identify hardcoded numbers that may not have an obvious meaning, promoting readability and maintainability.

- **Performance enhancement checks**: These include checks such as `cppcoreguidelines-avoid-c-arrays` and `cppcoreguidelines-avoid-non-const-global-variables`, which promote the use of modern C++ constructs such as `std::array` or `std::vector` over C-style arrays and discourage the use of non-const global variables.

- **Bounds safety checks**: Ensuring bounds safety is crucial in C++, and checks such as `cppcoreguidelines-pro-bounds-array-to-pointer-decay` and `cppcoreguidelines-pro-bounds-constant-array-index` warn against common pitfalls that can lead to **out-of-bounds (OOB)** errors.

- **Ownership and smart pointers checks**: These checks emphasize the correct usage of smart pointers to manage resource ownership. `cppcoreguidelines-owning-memory` guides developers on when and how to use smart pointers such as `std::unique_ptr` or `std::shared_ptr`.

- **Rule of Five and Rule of Zero checks**: Clang-Tidy enforces the Rule of Five and Rule of Zero in C++ class design, ensuring that classes managing resources correctly implement copy and move constructors/assignment operators or avoid managing resources manually, respectively.

- **Miscellaneous checks**: Other miscellaneous checks include `cppcoreguidelines-special-member-functions` (ensuring the correct implementation of special member functions) and `cppcoreguidelines-interfaces-global-init` (avoiding global initialization order issues).

  Adherence to the C++ Core Guidelines via Clang-Tidy checks can significantly improve the quality of C++ code, making it more robust, maintainable, and aligned with modern C++ practices. These checks cover a wide range of best practices and are generally considered good to follow for most C++ projects, especially those aiming to leverage modern C++ features effectively.

- **Check packages for standards compliance**: Clang-Tidy offers "packages" of checks that help ensure compliance with certain high-level standards:

  - **High-performance C++ (hi-cpp)**: These checks focus on ensuring that the code is optimized for performance.

  - **Certifications**: For projects that require adherence to specific certification standards (such as MISRA, CERT, and so on), Clang-Tidy offers checks that help align the code with these standards, although it's important to note that using these checks alone may not be sufficient for full compliance with such certifications.

The ability to add custom checks means that Clang-Tidy is not just a static analysis tool but a platform that can adapt to various coding standards and practices. This adaptability makes it an ideal choice for projects ranging from open source libraries to commercial software, each with its unique set of requirements and standards. By leveraging these specialized checks, teams can ensure that their code not only adheres to general best practices in C++ but also aligns with specific guidelines and nuances of their project or organization.

Fine-tuning Clang-Tidy for customized static analysis

Configuring Clang-Tidy effectively is key to harnessing its full potential in a C++ project. This involves not just enabling and disabling certain checks but also controlling how specific parts of the code are analyzed. By customizing its behavior, developers can ensure that Clang-Tidy's output is both relevant and actionable, focusing on the most important aspects of their code base. Let's take a closer look at this:

- **Enabling and disabling checks**: Use the `--checks=` option to enable specific checks, and prepend `-` to disable others. For instance, to turn on performance checks while turning off a specific one, you might use the following:

  ```
  clang-tidy my_code.cpp --checks='performance-*, -performance-
  noexcept-move-constructor'
  ```

- **Ignoring specific warnings**: Clang-Tidy allows the suppression of warnings in two primary ways:

  - **General suppression with NOLINT**: You can use the `NOLINT` comment to suppress all warnings for a specific line of code. This is a broad approach and might hide more than intended:

    ```
    int x = 0; // NOLINT
    ```

  - **Specific suppression**: A more targeted approach is to use `NOLINT(check-name)` to suppress specific warnings. This approach is preferable as it prevents over-suppression of potentially useful warnings:

    ```
    int x = 0; // NOLINT(bugprone-integer-division)
    ```

- **Treating warnings as errors**: To enforce code quality rigorously, convert warnings to errors using the `--warnings-as-errors=` option. This can be applied globally or to specific checks:

  ```
  clang-tidy my_code.cpp --warnings-as-errors='bugprone-*'
  ```

- **Using a configuration file**: For consistent and shared settings across a project, place a `.clang-tidy` file at the project's root. This file should list enabled checks and other configurations, as in the following example:

  ```
  Checks: 'performance-*, -performance-noexcept-move-constructor'
    WarningsAsErrors: 'bugprone-*'
  ```

Proper configuration of Clang-Tidy is crucial for effective static analysis in C++. By selectively enabling checks, specifically suppressing warnings where necessary, and treating critical warnings as errors, teams can maintain high code quality standards. The ability to fine-tune the analysis on a line-by-line basis with specific suppression comments ensures that Clang-Tidy provides focused and relevant feedback, making it an invaluable tool in the C++ developer's toolkit.

Overview of static analysis tools – comparing PVS-Studio, SonarQube, and others to Clang-Tidy

Static analysis tools are essential for ensuring code quality and adherence to best practices. While Clang-Tidy is a prominent tool in this space, particularly for C++ projects, there are other significant tools such as PVS-Studio and SonarQube, each with its unique features and strengths. Let's compare these tools to Clang-Tidy and also mention a few other notable options.

PVS-Studio

Using PVS-Studio has the following benefits:

- **Focus**: PVS-Studio is renowned for its deep analysis capabilities, particularly in detecting potential bugs, security flaws, and compliance with coding standards.

- **Languages supported**: While Clang-Tidy is focused primarily on C and C++, PVS-Studio supports a broader range of languages, including C#, Java, and even mixed C/C++/C# code bases.

- **Integration and usage**: PVS-Studio can be integrated into various IDEs and **continuous integration (CI)** systems. It differs from Clang-Tidy in that it's a standalone tool, not tied to a specific compiler such as Clang.

- **Unique features**: It offers extensive checks for potential code vulnerabilities and is often praised for its detailed diagnostic messages.

SonarQube

SonarQube offers the following benefits:

- **Focus**: SonarQube offers a comprehensive suite of code quality checks, including bugs, code smells, and security vulnerabilities

- **Languages supported**: It supports a wide range of programming languages, making it a versatile choice for multi-language projects

- **Integration and usage**: SonarQube stands out with its web-based dashboard that provides a detailed overview of the code quality, offering a more holistic view compared to Clang-Tidy

- **Unique features**: It includes features for code coverage and technical debt estimation, which are not the primary focus of Clang-Tidy

Other notable tools

Other notable tools in this field include the following:

- **Cppcheck**: Focused specifically on C and C++ languages, Cppcheck is a static analysis tool that emphasizes detecting undefined behavior, dangerous coding constructs, and other subtle bugs that other tools might miss. It's lightweight and can complement Clang-Tidy well.

- **Coverity**: Known for its high accuracy and support for a wide range of programming languages, Coverity is a tool used in both commercial and open source projects to detect software defects and security vulnerabilities.

- **Visual Studio Static Analysis**: Integrated into the Visual Studio IDE, this tool provides checks specifically tailored for Windows development. It's highly useful for developers working primarily in the Windows ecosystem.

Comparison with Clang-Tidy

Let's now compare the aforementioned tools with Clang-Tidy:

- **Language support**: Clang-Tidy is primarily focused on C and C++, while tools such as PVS-Studio, SonarQube, and Coverity support a broader range of languages.

- **Integration and reporting**: Clang-Tidy is closely integrated with the LLVM/Clang ecosystem, making it an excellent choice for projects already using these tools. In contrast, SonarQube offers a comprehensive dashboard and PVS-Studio provides detailed reports, which are beneficial for larger projects or teams.

- **Specific use cases**: Tools such as Cppcheck and Visual Studio Static Analysis have specific niches – Cppcheck for its lightweight nature and focus on C/C++, and Visual Studio Static Analysis for its Windows-specific checks.

- **Commercial versus open source**: Clang-Tidy is open source and free to use, whereas tools such as Coverity and PVS-Studio are commercial products, offering enterprise-level features and support.

Summary

In this chapter, we took a deep dive into the world of static analysis for C++ development, exploring a variety of tools and methodologies. We began with an overview of Clang-Tidy, developed by the LLVM Foundation, and its extensive capabilities in checking code for performance issues, modernization, bugs, style, readability, and security. We also covered other significant tools in the static analysis domain, including PVS-Studio, known for its vulnerability detection and multi-language support; SonarQube, with its comprehensive code quality checks and intuitive dashboard; and others such as Cppcheck, Coverity, and Visual Studio Static Analysis, each bringing unique strengths to the table.

A significant focus was on configuring Clang-Tidy, detailing how to fine-tune it for specific project needs, such as enabling or disabling diagnostics, managing warnings, and setting up configuration files. We also discussed the tool's extensibility, highlighting custom checks for different coding standards and compliance packages for various requirements such as high-performance C++ and certifications.

This exploration provided us with a broader understanding of the static analysis landscape in C++, revealing how these tools can significantly enhance code quality. As we wrap up this chapter, we prepare to shift gears in the next one, where we will explore the realm of dynamic analysis, complementing our knowledge of static analysis with insights into the behavior of running code. This will complete our comprehensive look at tools and techniques essential for mastering C++ code quality.

11
Dynamic Analysis

In the intricate world of software development, ensuring the correctness, efficiency, and security of code is not just a goal but a necessity. This is particularly true in C++ programming, where the power and complexity of the language present both opportunities and challenges. One of the most effective approaches to maintaining high code quality in C++ is **dynamic code analysis** – a process that scrutinizes program behavior as it runs to detect a range of potential issues.

Dynamic code analysis stands in contrast to static analysis, which examines source code without executing it. While static analysis is invaluable for catching syntax errors, code smells, and certain types of bugs early in the development cycle, dynamic analysis delves deeper. It uncovers issues that only manifest during the actual execution of the program, such as memory leaks, race conditions, and other runtime errors that can lead to crashes, erratic behavior, or security vulnerabilities.

This chapter aims to explore the landscape of dynamic code analysis tools in C++, with a particular focus on some of the most powerful and widely used tools in the industry: a suite of compiler-based sanitizers, including **AddressSanitizer (ASan)**, **ThreadSanitizer (TSan)**, and **UndefinedBehaviorSanitizer (UBSan)**, as well as Valgrind, a versatile tool known for its thorough memory debugging capabilities.

Compiler sanitizers, part of the LLVM project and GCC project, offer a range of options for dynamic analysis. ASan is remarkable for its ability to detect various memory-related errors, TSan excels in identifying race conditions in multi-threaded code, and UBSan helps in catching undefined behaviors that can lead to unpredictable program behavior. These tools are praised for their efficiency, precision, and ease of integration into existing development workflows. Most of them are supported by GCC and MSVC as well.

On the other hand, Valgrind, an instrumentation framework for building dynamic analysis tools, shines with its comprehensive memory leak detection and the ability to analyze binary executables without requiring source code recompilation. It's a go-to solution for complex scenarios where in-depth memory analysis is paramount, albeit at the cost of higher performance overhead.

Throughout this chapter, we will delve into each of these tools, understanding their strengths, weaknesses, and appropriate use cases. We'll explore how they can be effectively integrated into your C++ development process, and how they complement each other to provide a robust framework for ensuring the quality and reliability of C++ applications.

By the end of this chapter, you will have a thorough understanding of dynamic code analysis in C++, equipped with the knowledge to choose and utilize the right tools for your specific development needs, ultimately leading to cleaner, more efficient, and reliable C++ code.

Compiler-based dynamic code analysis

Compiler-based sanitizers contain two parts: compiler instrumentation and runtime diagnostics:

- **Compiler instrumentation**: When you compile your C++ code with sanitizers enabled, the compiler instruments the generated binary with additional checks. These checks are strategically inserted into the code to monitor for specific types of errors. For instance, ASan adds code to track memory allocations and accesses, enabling it to detect memory misuses such as buffer overflows and memory leaks.

- **Runtime diagnostics**: As the instrumented program runs, these checks actively monitor the program's behavior. When a sanitizer detects an error (such as a memory access violation or a data race), it immediately reports this, often with detailed information about the location and nature of the error. This real-time feedback is invaluable for identifying and fixing elusive bugs that might be difficult to catch through traditional testing.

Despite all compiler teams constantly working on adding new sanitizers and improving the existing ones, there are still some limitations to the compiler-based sanitizers:

- **Clang and GCC**: Most sanitizers, including ASan, TSan, and UBSan, are supported by both Clang and GCC. This wide support makes them accessible to a large portion of the C++ development community, regardless of the preferred compiler.

- **Microsoft Visual C++ (MSVC)**: MSVC also supports some sanitizers, though the range and capabilities might differ from those in Clang and GCC. For example, MSVC has support for ASan, which is useful for Windows-specific C++ development.

- **Cross-platform utility**: The cross-compiler and cross-platform nature of these tools mean they can be used in a variety of development environments, from Linux and macOS to Windows, enhancing their utility in diverse C++ projects.

ASan

ASan is a runtime memory error detector, part of the LLVM compiler infrastructure, GCC, and MSVC. It serves as a specialized tool for developers to identify and resolve various kinds of memory-related errors, including, but not limited to, buffer overflows, dangling pointer accesses, and memory leaks.

The tool achieves this by instrumenting the code during the compilation process, enabling it to monitor memory accesses and allocations at runtime.

One of the key strengths of ASan is its ability to provide detailed error reports. When a memory error is detected, ASan outputs comprehensive information, including the type of error, the memory location involved, and the stack trace. This level of detail significantly aids in the debugging process, allowing developers to pinpoint the source of the issue quickly.

Integrating ASan into a C++ development workflow is straightforward. It requires minimal changes to the build process, typically involving the addition of a compiler flag (`-fsanitize=address`) during compilation. For better results, it makes sense to use reasonable performance add `-O1` or higher. To get nicer stack traces in error messages, add `-fno-omit-frame-pointer`. This ease of integration, combined with its effectiveness in catching memory errors, makes ASan an indispensable tool for developers aiming to enhance the reliability and security of their C++ applications.

Symbolizing reports in ASan

When using ASan to detect memory errors in C++ applications, it's crucial to symbolize the error reports. Symbolization translates memory addresses and offsets in ASan's output into human-readable function names, file names, and line numbers. This process is vital for effective debugging, as it allows developers to easily identify where in the source code the memory error occurred.

Without symbolization, the ASan report provides less meaningful raw memory addresses, making it challenging to trace back to the exact location in the source code where the error happened. Symbolized reports, on the other hand, offer clear and actionable insights, enabling developers to quickly understand and fix the underlying issues in their code.

The configuration of ASan symbolization is typically automatic, requiring no additional steps. However, in some cases, you might need to explicitly set the `ASAN_SYMBOLIZER_PATH` environment variable to point to the symbolizer tool. This is especially true on non-Linux Unix systems, where additional tools such as `addr2line` might be required for symbolization. If it does not work out of the box, please go over the following steps to ensure that symbolization is configured correctly:

1. **Compile with debug information**:

    *   Ensure that your program is compiled with debug information. This is typically done by adding the `-g` flag to your compilation command. For instance:

        ```
        clang++ -fsanitize=address -g -o your_program your_file.cpp
        ```

    *   Compiling with `-g` includes debugging symbols in the binary, which are essential for symbolization.

2. **Use an ASan symbolizer**:

    *   ASan typically autoconfigures its symbolizer if the required tools are available on your system. This means symbolization often works out of the box.

- On Unix-like systems, ASan uses the LLVM symbolizer by default. Ensure that the `llvm-symbolizer` tool is in your system's PATH.

3. **Check for additional tools (optional):**

- On some systems, especially non-Linux Unix systems, you might need additional tools for symbolization.

- Tools such as `addr2line` (part of GNU Binutils) can be used for symbolizing stack traces.

4. **Environment variables (if needed):**

- In cases where automatic symbolization doesn't work, you can set the `ASAN_SYMBOLIZER_PATH` environment variable to point to the symbolizer tool. For example:

```
export ASAN_SYMBOLIZER_PATH=/path/to/llvm-symbolizer
```

- This explicitly tells ASan which symbolizer to use.

5. **Running your program:**

- Run your compiled program as usual. If a memory error is detected, ASan will output a symbolized stack trace.

- The report will include function names, file names, and line numbers, making it easier to locate and address the error in your code.

Out-of-bounds access

Let us try to catch one of the most critical errors of C++ programming: **out-of-bounds access**. This issue spans various segments of memory management – the heap, the stack, and global variables, each presenting unique challenges and risks.

Out-of-bounds access on the heap

We begin by exploring out-of-bounds access on the heap, where dynamic memory allocation can lead to pointers exceeding the allocated memory boundaries. Consider the following example:

```
int main() {
    int *heapArray = new int[5];
    heapArray[5]   = 10; // Out-of-bounds write on the heap
    delete[] heapArray;
    return 0;
}
```

This code snippet demonstrates an out-of-bounds write, attempting to access an index that is beyond the allocated range, leading to undefined behavior and potential memory corruption.

If we run this code with ASan enabled, we get the following output:

```
make && ./a.out
================================================================
==3102850==ERROR: AddressSanitizer: heap-buffer-overflow on
address 0x603000000054 at pc 0x55af5525f222 bp 0x7ffde596fb60 sp
0x7ffde596fb50
WRITE of size 4 at 0x603000000054 thread T0
    #0 0x55af5525f221 in main /home/user/clang-sanitizers/main.cpp:3
    #1 0x7f1ad0a29d8f  (/lib/x86_64-linux-gnu/libc.so.6+0x29d8f)
    #2 0x7f1ad0a29e3f in __libc_start_main (/lib/x86_64-linux-gnu/
libc.so.6+0x29e3f)
    #3 0x55af5525f104 in _start (/home/user/clang-sanitizers/build/a.
out+0x1104)

0x603000000054 is located 0 bytes to the right of 20-byte region
[0x603000000040,0x603000000054)
allocated by thread T0 here:
    #0 0x7f1ad12b6357 in operator new[](unsigned long) ../../../../
src/libsanitizer/asan/asan_new_delete.cpp:102
    #1 0x55af5525f1de in main /home/user/clang-sanitizers/main.cpp:2
    #2 0x7f1ad0a29d8f  (/lib/x86_64-linux-gnu/libc.so.6+0x29d8f)

SUMMARY: AddressSanitizer: heap-buffer-overflow /home/user/clang-
sanitizers/main.cpp:3 in main
Shadow bytes around the buggy address:
  0x0c067fff7fb0: 00 00 00 00 00 00 00 00 00 00 00 00 00 00 00 00
  0x0c067fff7fc0: 00 00 00 00 00 00 00 00 00 00 00 00 00 00 00 00
  0x0c067fff7fd0: 00 00 00 00 00 00 00 00 00 00 00 00 00 00 00 00
  0x0c067fff7fe0: 00 00 00 00 00 00 00 00 00 00 00 00 00 00 00 00
  0x0c067fff7ff0: 00 00 00 00 00 00 00 00 00 00 00 00 00 00 00 00
=>0x0c067fff8000: fa fa 00 00 00 fa fa fa 00 00[04]fa fa fa fa fa
  0x0c067fff8010: fa fa fa fa fa fa fa fa fa fa fa fa fa fa fa fa
  0x0c067fff8020: fa fa fa fa fa fa fa fa fa fa fa fa fa fa fa fa
  0x0c067fff8030: fa fa fa fa fa fa fa fa fa fa fa fa fa fa fa fa
  0x0c067fff8040: fa fa fa fa fa fa fa fa fa fa fa fa fa fa fa fa
  0x0c067fff8050: fa fa fa fa fa fa fa fa fa fa fa fa fa fa fa fa
Shadow byte legend (one shadow byte represents 8 application bytes):
  Addressable:           00
  Partially addressable: 01 02 03 04 05 06 07
  Heap left redzone:       fa
  Freed heap region:       fd
  Stack left redzone:      f1
  Stack mid redzone:       f2
  Stack right redzone:     f3
```

```
Stack after return:        f5
Stack use after scope:     f8
Global redzone:            f9
Global init order:         f6
Poisoned by user:          f7
Container overflow:        fc
Array cookie:              ac
Intra object redzone:      bb
ASan internal:             fe
Left alloca redzone:       ca
Right alloca redzone:      cb
Shadow gap:                cc
==3102850==ABORTING
```

As you can see, the report includes a detailed stack trace, highlighting the exact location of the error in the source code. This information is invaluable for debugging and fixing the issue.

Out-of-bounds access on the stack

Next, we focus on the stack. Here, out-of-bounds accesses often occur with local variables due to incorrect indexing or buffer overruns. For example:

```
int main() {
    int stackArray[5];
    stackArray[5] = 10; // Out-of-bounds write on the stack
    return 0;
}
```

In this case, accessing `stackArray[5]` is out of bounds, as valid indices are from 0 to 4. Such errors can result in crashes or exploitable vulnerabilities. The output of ASan for this example is very similar to the previous one:

```
==3190568==ERROR: AddressSanitizer: stack-buffer-overflow on
address 0x7ffd166961e4 at pc 0x55b4cd113295 bp 0x7ffd166961a0 sp
0x7ffd16696190
WRITE of size 4 at 0x7ffd166961e4 thread T0
    #0 0x55b4cd113294 in main /home/user/clang-sanitizers/main.cpp:3
    #1 0x7f90fc829d8f  (/lib/x86_64-linux-gnu/libc.so.6+0x29d8f)
    #2 0x7f90fc829e3f in __libc_start_main (/lib/x86_64-linux-gnu/
libc.so.6+0x29e3f)
    #3 0x55b4cd113104 in _start (/home/user/clang-sanitizers/build/a.
out+0x1104)

Address 0x7ffd166961e4 is located in stack of thread T0 at offset 52
in frame
```

```
    #0 0x55b4cd1131d8 in main /home/user/clang-sanitizers/main.cpp:1
```

This frame has 1 object(s):
    [32, 52) 'stackArray' (line 2) <== Memory access at offset 52
overflows this variable
HINT: this may be a false positive if your program uses some custom
stack unwind mechanism, swapcontext or vfork
      (longjmp and C++ exceptions *are* supported)
SUMMARY: AddressSanitizer: stack-buffer-overflow /home/user/clang-
sanitizers/main.cpp:3 in main
Shadow bytes around the buggy address:
```
  0x100022ccabe0: 00 00 00 00 00 00 00 00 00 00 00 00 00 00 00 00
  0x100022ccabf0: 00 00 00 00 00 00 00 00 00 00 00 00 00 00 00 00
  0x100022ccac00: 00 00 00 00 00 00 00 00 00 00 00 00 00 00 00 00
  0x100022ccac10: 00 00 00 00 00 00 00 00 00 00 00 00 00 00 00 00
  0x100022ccac20: 00 00 00 00 00 00 00 00 00 00 00 00 00 00 00 00
=>0x100022ccac30: 00 00 00 00 00 00 f1 f1 f1 f1 00 00[04]f3 f3 f3
  0x100022ccac40: f3 f3 00 00 00 00 00 00 00 00 00 00 00 00 00 00
  0x100022ccac50: 00 00 00 00 00 00 00 00 00 00 00 00 00 00 00 00
  0x100022ccac60: 00 00 00 00 00 00 00 00 00 00 00 00 00 00 00 00
  0x100022ccac70: 00 00 00 00 00 00 00 00 00 00 00 00 00 00 00 00
  0x100022ccac80: 00 00 00 00 00 00 00 00 00 00 00 00 00 00 00 00
```
Shadow byte legend (one shadow byte represents 8 application bytes):
```
  Addressable:           00
  Partially addressable: 01 02 03 04 05 06 07
  Heap left redzone:       fa
  Freed heap region:       fd
  Stack left redzone:      f1
  Stack mid redzone:       f2
  Stack right redzone:     f3
  Stack after return:      f5
  Stack use after scope:   f8
  Global redzone:          f9
  Global init order:       f6
  Poisoned by user:        f7
  Container overflow:      fc
  Array cookie:            ac
  Intra object redzone:    bb
  ASan internal:           fe
  Left alloca redzone:     ca
  Right alloca redzone:    cb
  Shadow gap:              cc
==3190568==ABORTING
```

Out-of-bounds access to global variables

Finally, we examine global variables. These are susceptible to similar risks when accessed beyond their defined boundaries:

```
int globalArray[5];

int main() {
    globalArray[5] = 10;   // Out-of-bounds access to a global array
    return 0;
}
```

Here, the attempt to write to `globalArray[5]` is an out-of-bounds operation, leading to undefined behavior. Since the output of ASan is similar to the previous examples, we won't include it here.

Addressing use-after-free vulnerabilities in C++

In the following section, we will address a critical and often challenging issue in C++ programming: **use-after-free vulnerabilities**. This type of error occurs when a program continues to use a memory location after it has been freed, leading to undefined behavior, program crashes, security vulnerabilities, and data corruption. We'll explore this issue in various contexts, providing insights into its identification and prevention.

Use-after-free in dynamic memory (heap)

The most common scenario for use-after-free errors occurs with dynamically allocated memory on the heap. Consider the following example:

```
#include <iostream>

template <typename T>
struct Node {
    T data;
    Node *next;
    Node(T val) : data(val), next(nullptr) {}
};

int main() {
    auto *head = new Node(1);
    auto *temp = head;
    head       = head->next;
    delete temp;
    std::cout << temp->data; // Use-after-free in a linked list
```

```
    return 0;
}
```

In this snippet, the memory pointed to by `ptr` is accessed after it has been freed with `delete`. This access can lead to unpredictable behavior, as the freed memory might be allocated for other purposes or modified by the system.

Use-after-free with object references

Use-after-free can also occur in object-oriented programming, especially when dealing with references or pointers to objects that have been destroyed. For instance:

```
class Example {
public:
    int value;
    Example() : value(0) {}
};

Example* obj = new Example();
Example& ref = *obj;
delete obj;
std::cout << ref.value;  // Use-after-free through a reference
```

Here, `ref` refers to an object that has been deleted, and any operation on `ref` after the deletion leads to use-after-free.

Use-After-Free in Complex Data Structures

Complex data structures, such as linked lists or trees, are also prone to use-after-free errors, particularly during deletion or restructuring operations. For example:

```
struct Node {
    int data;
    Node* next;
    Node(int val) : data(val), next(nullptr) {}
};

Node* head = new Node(1);
Node* temp = head;
head = head->next;
delete temp;
std::cout << temp->data;  // Use-after-free in a linked list
```

In this case, `temp` is used after it has been freed, which can lead to serious issues, especially if the list is large or part of a critical system component.

ASan can help in detecting use-after-free errors in C++ programs. For instance, if we run the previous example with ASan enabled, we get the following output:

```
make && ./a.out
Consolidate compiler generated dependencies of target a.out
[100%] Built target a.out
===================================================================
==3448347==ERROR: AddressSanitizer: heap-use-after-free on
address 0x602000000010 at pc 0x55fbcc2ca3b2 bp 0x7fff2f3af7a0 sp
0x7fff2f3af790
READ of size 4 at 0x602000000010 thread T0
    #0 0x55fbcc2ca3b1 in main /home/user/clang-sanitizers/main.cpp:15
    #1 0x7efdb6429d8f  (/lib/x86_64-linux-gnu/libc.so.6+0x29d8f)
    #2 0x7efdb6429e3f in __libc_start_main (/lib/x86_64-linux-gnu/
libc.so.6+0x29e3f)
    #3 0x55fbcc2ca244 in _start (/home/user/clang-sanitizers/build/a.
out+0x1244)
```

Use-after-return detection in ASan

Use-after-return is a type of memory error in C++ programming where a function returns a pointer or a reference to a local (stack-allocated) variable. This local variable ceases to exist once the function returns, making any subsequent access through the returned pointer or reference invalid and dangerous. This can lead to undefined behavior and potential security vulnerabilities.

ASan provides a mechanism to detect use-after-return errors. It can be controlled using the `-fsanitize-address-use-after-return` flag during compilation and the `ASAN_OPTIONS` environment variable at runtime.

The following describes the configuration of use-after-return detection:

- **Compilation flag**: `-fsanitize-address-use-after-return=(never|runtime|always)`

  The flag accepts three settings:

  - `never`: This disables use-after-return detection
  - `runtime`: This enables detection, but it can be overridden at runtime (default setting)
  - `always`: This always enables detection, irrespective of runtime settings

- **Runtime configuration**:

  To explicitly enable or disable use-after-return detection at runtime, use the ASAN_OPTIONS environment variable:

  - **Enable**: ASAN_OPTIONS=detect_stack_use_after_return=1
  - **Disable**: ASAN_OPTIONS=detect_stack_use_after_return=0
  - On Linux, detection is enabled by default

Here is an example of its usage:

1. **Compiling with use-after-return detection enabled**:

   ```
   clang++ -fsanitize=address -fsanitize-address-use-after-
   return=always -g -o your_program your_file.cpp
   ```

   This command compiles your_file.cpp with ASan and explicitly enables use-after-return detection.

2. **Running with detection enabled/disabled**:

   - To run the program with use-after-return detection enabled (on platforms where it's not the default):

     ```
     ASAN_OPTIONS=detect_stack_use_after_return=1 ./your_program
     ```

   - To disable detection, even if it was enabled at compile time:

     ```
     ASAN_OPTIONS=detect_stack_use_after_return=0 ./your_program
     ```

Example code demonstrating use-after-return

The provided C++ code example demonstrates a use-after-return scenario, which is a type of undefined behavior caused by returning a reference to a local variable from a function. Let's analyze the example and understand the implications:

```cpp
#include <iostream>

const std::string &get_binary_name() {
    const std::string name = "main";
    return name; // Returning address of a local variable
}

int main() {
    const auto &name = get_binary_name();
    // Use after return: accessing memory through name is undefined
behavior
```

```
    std::cout << name << std::endl;
    return 0;
}
```

In the given code example, the get_binary_name function is designed to create a local std::string object named name and return a reference to it. The critical issue arises from the fact that name is a local variable, which gets destroyed as soon as the function scope ends. As a result, the reference that get_binary_name returns becomes invalid the moment the function exits.

In the main function, this returned reference, now stored in name, is used to access the string value. However, since name refers to a local variable that has already been destroyed, using it in this manner leads to undefined behavior. This is a classic example of a use-after-return error, where the program attempts to access memory that is no longer valid.

The function's intended functionality seems to be to return the program's name. However, for this to work correctly, the name variable should have a static or global lifetime rather than being a local variable confined to the get_binary_name function. This would ensure that the returned reference remains valid beyond the scope of the function, avoiding the use-after-return error.

Modern compilers are equipped with the ability to issue warnings about potentially problematic code patterns, such as returning references to local variables. In the context of our example, a compiler might flag the return of a local variable reference as a warning, signaling a possible use-after-return error.

However, to effectively demonstrate the capabilities of ASan in catching use-after-return errors, it's sometimes necessary to bypass these compile-time warnings. This can be achieved by explicitly disabling the compiler's warning. For instance, by adding the -Wno-return-local-addr flag to the compilation command, we can prevent the compiler from issuing a warning about returning a local address. Doing so allows us to shift the focus from compile-time detection to runtime detection, where ASan's capabilities in identifying use-after-return errors can be more prominently displayed and tested. This approach underscores the runtime diagnostic strengths of ASan, particularly in cases where compile-time analysis might not be sufficient.

Compiling with ASan

To compile this program with ASan's use-after-return detection enabled, you would use a command such as the following:

```
clang++ -fsanitize=address -Wno-return-local-addr -g your_file.cpp -o
your_program
```

This command compiles the program with ASan enabled while suppressing the specific compiler warning about returning the address of a local variable. Running the compiled program will allow ASan to detect and report the use-after-return error at runtime:

```
Consolidate compiler generated dependencies of target a.out
[100%] Built target a.out
```

```
AddressSanitizer:DEADLYSIGNAL
=================================================================
==4104819==ERROR: AddressSanitizer: SEGV on unknown address
0x000000000008 (pc 0x7f74e354f4c4 bp 0x7ffefcd298e0 sp 0x7ffefcd298c8
T0)
==4104819==The signal is caused by a READ memory access.
==4104819==Hint: address points to the zero page.
    #0 0x7f74e354f4c4 in std::basic_ostream<char, std::char_
traits<char> >& std::operator<< <char, std::char_traits<char>,
std::allocator<char> >(std::basic_ostream<char, std::char_traits<char>
>&, std::__cxx11::basic_string<char, std::char_traits<char>,
std::allocator<char> > const&) (/lib/x86_64-linux-gnu/libstdc++.
so.6+0x14f4c4)
    #1 0x559799ab4785 in main /home/user/clang-sanitizers/main.cpp:11
    #2 0x7f74e3029d8f  (/lib/x86_64-linux-gnu/libc.so.6+0x29d8f)
    #3 0x7f74e3029e3f in __libc_start_main (/lib/x86_64-linux-gnu/
libc.so.6+0x29e3f)
    #4 0x559799ab4504 in _start (/home/user/clang-sanitizers/build/a.
out+0x2504)

AddressSanitizer can not provide additional info.
SUMMARY: AddressSanitizer: SEGV (/lib/x86_64-linux-gnu/libstdc++.
so.6+0x14f4c4) in std::basic_ostream<char, std::char_traits<char> >&
std::operator<< <char, std::char_traits<char>, std::allocator<char>
>(std::basic_ostream<char, std::char_traits<char> >&, std::__
cxx11::basic_string<char, std::char_traits<char>, std::allocator<char>
> const&)
==4104819==ABORTING
```

This example highlights the importance of understanding object lifetimes in C++ and how misuse can lead to undefined behavior. While compiler warnings are valuable for catching such issues at compile time, tools such as ASan provide an additional layer of runtime error detection, which is especially useful in complex scenarios where compile-time analysis might not suffice.

Use-after-return detection

The concept of use-after-scope in C++ involves accessing a variable after its scope has ended, leading to undefined behavior. This type of error is subtle and can be particularly challenging to detect and debug. ASan offers a feature to detect use-after-scope errors, which can be enabled using the -fsanitize-address-use-after-scope compilation flag.

Understanding use-after-scope

Use-after-scope occurs when a program continues to use a pointer or reference to a variable that has gone out of scope. Unlike use-after-return, where the issue is with function-local variables, use-after-scope can occur within any scope, such as within a block of code, such as an if statement or a loop.

When a variable goes out of scope, its memory location may still hold the old data for some time, but this memory can be overwritten at any moment. Accessing this memory is undefined behavior and can lead to erratic program behavior or crashes.

Configuring ASan for use-after-scope detection

Compilation flag: `-fsanitize-address-use-after-scope`:

- Adding this flag to your compilation command instructs ASan to instrument the code to detect use-after-scope errors

- It's important to note that this detection is not enabled by default and must be explicitly enabled

Example code demonstrating use-after-scope

The provided code snippet demonstrates a classic case of use-after-scope error in C++. Let's analyze the code and understand the issue:

```
int* create_array(bool condition) {
  int *p;
  if (condition) {
    int x[10];
    p = x;
  }
  *p = 1;
}
```

In the given code snippet, we begin by declaring a p pointer without initializing it. The function then enters a conditional scope where, if `condition` is true, an `x[10]` array is created on the stack. Within this scope, the p pointer is assigned to point to the start of this array, effectively making p point to x.

The critical issue arises after the conditional block is exited. At this point, the x array, being local to the `if` block, goes out of scope and is no longer valid. However, the p pointer still holds the address of where x was located. When the code attempts to write to this memory location using `*p = 1;`, it is trying to access the memory of the now out-of-scope x array. This action leads to a use-after-scope error, where p is dereferenced to access memory that is no longer valid within the current scope. This kind of error is a classic example of use-after-scope, highlighting the dangers of accessing memory through pointers that point to out-of-scope variables.

Accessing memory through a pointer that points to an out-of-scope variable, as demonstrated in the provided code snippet, leads to undefined behavior. This is because once the x variable goes out of scope, the memory location to which p points becomes indeterminate. The undefined behavior arising from this scenario is problematic for several reasons.

Firstly, it poses significant security and stability risks to the program. The undefined nature of the behavior means that the program could crash or behave unpredictably. Such instability in a program's execution can have far-reaching consequences, particularly in applications where reliability is critical. Furthermore, if the memory location previously occupied by x gets overwritten by other parts of the program, it could potentially lead to security vulnerabilities. These vulnerabilities might be exploited to compromise the program or the system on which it is running.

In summary, the undefined behavior resulting from accessing memory through pointers to out-of-scope variables is a serious concern in software development, necessitating careful management of variable scope and memory access patterns to ensure the security and stability of the program.

To compile the program with ASan's use-after-scope detection enabled, you would use a command such as the following:

```
g++ -fsanitize=address -fsanitize-address-use-after-scope -g your_
file.cpp -o your_program
```

Running the compiled program with these settings enables ASan to detect and report use-after-scope errors at runtime.

Use-after-scope errors can be insidious and difficult to trace due to their dependence on the program's runtime state and memory layout. By enabling use-after-scope detection in ASan, developers gain a valuable tool for identifying these errors, leading to more robust and reliable C++ applications. Understanding and preventing such issues is crucial for writing safe and correct C++ code.

Double-free and invalid-free checks in ASan

ASan, a part of the LLVM project, provides robust mechanisms to detect and diagnose two critical types of memory errors in C++ programs: double free and invalid free. These errors are not only common in complex C++ applications but can also lead to severe issues such as program crashes, undefined behavior, and security vulnerabilities.

Understanding double free and invalid free

Understanding double-free and invalid-free errors is essential in managing memory in C++ programs effectively.

A double-free error occurs when an attempt is made to free a memory block more than once using the delete or delete[] operators. This typically happens when the same memory allocation is passed to delete or delete[] twice. The first call to delete frees the memory, but the second call attempts to free memory that has already been released. This can lead to heap corruption, as the program might subsequently modify or reallocate the freed memory for other uses. Double-free errors can cause unpredictable behavior in your program, including crashes and data corruption.

Invalid-free errors, on the other hand, occur when delete or delete[] is used on a pointer that wasn't allocated with new or new[], or that has already been freed. This category includes attempts to free a null pointer, pointers to stack memory (which are not dynamically allocated), or pointers to uninitialized memory. Like double-free errors, invalid frees can also lead to heap corruption and unpredictable program behavior. They are particularly insidious because they can corrupt the memory management structures of the C++ runtime, leading to subtle and hard-to-diagnose bugs.

Both of these errors stem from improper handling of dynamic memory, underscoring the importance of adhering to best practices in memory management, such as ensuring every new has a corresponding delete and avoiding the reuse of pointers after they have been freed.

This list outlines the features of ASan's detection mechanism:

- **Heap corruption detection**: ASan instruments the program to keep track of all heap allocations and deallocations. When a delete operation is performed, ASan checks whether the pointer corresponds to a valid, previously allocated, and not-yet-freed memory block.

- **Error reporting**: If a double-free or invalid-free error is detected, ASan aborts the program's execution and provides a detailed error report. This report includes the location in the code where the error occurred, the memory address involved, and the allocation history of that memory (if available).

Here is some example code demonstrating double-free errors:

```
int main() {
    int* ptr = new int(10);
    delete ptr;
    delete ptr;  // Double-free error
    return 0;
}
```

ASan would report the following error:

```
make && ./a.out
Consolidate compiler generated dependencies of target a.out
[ 50%] Building CXX object CMakeFiles/a.out.dir/main.cpp.o
[100%] Linking CXX executable a.out
[100%] Built target a.out
=====================================================================
==765374==ERROR: AddressSanitizer: attempting double-free on
0x602000000010 in thread T0:
    #0 0x7f7ff5eb724f in operator delete(void*, unsigned long)
../../../../src/libsanitizer/asan/asan_new_delete.cpp:172
    #1 0x55839eca830b in main /home/user/clang-sanitizers/main.cpp:6
    #2 0x7f7ff5629d8f  (/lib/x86_64-linux-gnu/libc.so.6+0x29d8f)
    #3 0x7f7ff5629e3f in __libc_start_main (/lib/x86_64-linux-gnu/
libc.so.6+0x29e3f)
    #4 0x55839eca81c4 in _start (/home/user/clang-sanitizers/build/a.
out+0x11c4)
```

```
0x602000000010 is located 0 bytes inside of 4-byte region
[0x602000000010,0x602000000014)
freed by thread T0 here:
    #0 0x7f7ff5eb724f in operator delete(void*, unsigned long)
../../../../src/libsanitizer/asan/asan_new_delete.cpp:172
    #1 0x55839eca82f5 in main /home/user/clang-sanitizers/main.cpp:5
    #2 0x7f7ff5629d8f  (/lib/x86_64-linux-gnu/libc.so.6+0x29d8f)

previously allocated by thread T0 here:
    #0 0x7f7ff5eb61e7 in operator new(unsigned long) ../../../../src/
libsanitizer/asan/asan_new_delete.cpp:99
    #1 0x55839eca829e in main /home/user/clang-sanitizers/main.cpp:4
    #2 0x7f7ff5629d8f  (/lib/x86_64-linux-gnu/libc.so.6+0x29d8f)

SUMMARY: AddressSanitizer: double-free ../../../../src/libsanitizer/
asan/asan_new_delete.cpp:172 in operator delete(void*, unsigned long)
==765374==ABORTING
```

In this example, the same memory pointed to by `ptr` is freed twice, leading to a double-free error.

Example code demonstrating invalid free

The provided code snippet demonstrates an invalid-free error, which is a type of memory management mistake that can occur in C++ programming. Let's dissect the example to understand the issue and its implications:

```
int main() {
    int local_var = 42;
    int* ptr = &local_var;
    delete ptr;  // Invalid free error
    return 0;
}
```

In a given code segment, we start by declaring and initializing a local int `local_var = 42;` variable. This creates a stack-allocated integer variable named `local_var`. Following this, a pointer assignment is made with int* `ptr = &local_var;`, where the `ptr` pointer is set to point to the address of `local_var`. This establishes a link between the pointer and the stack-allocated variable.

However, an issue arises with the subsequent operation: `delete ptr;`. This line of code attempts to free the memory pointed to by `ptr`. The problem here is that `ptr` is pointing to a stack-allocated variable, `local_var`, rather than a dynamically allocated piece of memory from the heap. In C++, the `delete` operator is intended to be used exclusively with pointers that have been allocated with `new`. Since `local_var` was not allocated with `new` (being a stack-allocated variable), using `delete` on `ptr` is invalid and leads to undefined behavior. This misuse of the `delete` operator on a non-heap pointer is a common mistake that can lead to serious runtime errors in a C++ program.

Here are some modern compiler warnings:

- Modern C++ compilers typically issue warnings or errors when delete is used on a pointer that doesn't point to dynamically allocated memory, as this is a common source of bugs.

- To compile this code without modification and demonstrate ASan's ability to catch such errors, you might need to suppress the compiler warning. This can be done by adding the -Wno-free-nonheap-object flag to the compilation command.

Compiling with ASan for invalid-free detection

To compile the program with ASan to detect the invalid free operation, use the following command:

```
clang++ -fsanitize=address -Wno-free-nonheap-object -g your_file.cpp
-o your_program
```

This command compiles the program with ASan enabled and suppresses the specific compiler warning about freeing non-heap objects. When you run the compiled program, ASan will detect and report the invalid free operation:

```
=================================================================
==900629==ERROR: AddressSanitizer: attempting free on address which
was not malloc()-ed: 0x7fff390f21d0 in thread T0
    #0 0x7f30b82b724f in operator delete(void*, unsigned long)
../../../../src/libsanitizer/asan/asan_new_delete.cpp:172
    #1 0x563f21cd72c7 in main /home/user/clang-sanitizers/main.cpp:4
    #2 0x7f30b7a29d8f  (/lib/x86_64-linux-gnu/libc.so.6+0x29d8f)
    #3 0x7f30b7a29e3f in __libc_start_main (/lib/x86_64-linux-gnu/
libc.so.6+0x29e3f)
    #4 0x563f21cd7124 in _start (/home/user/clang-sanitizers/build/a.
out+0x1124)

Address 0x7fff390f21d0 is located in stack of thread T0 at offset 32
in frame
    #0 0x563f21cd71f8 in main /home/user/clang-sanitizers/main.cpp:1

  This frame has 1 object(s):
    [32, 36) 'local_var' (line 2) <== Memory access at offset 32 is
inside this variable
HINT: this may be a false positive if your program uses some custom
stack unwind mechanism, swapcontext or vfork
    (longjmp and C++ exceptions *are* supported)
SUMMARY: AddressSanitizer: bad-free ../../../../src/libsanitizer/asan/
asan_new_delete.cpp:172 in operator delete(void*, unsigned long)
==900629==ABORTING
```

The attempt to delete a pointer to a non-heap object, as shown in the example, is a misuse of memory management operations in C++. Such practices can lead to undefined behavior and can potentially cause crashes or other erratic program behavior. ASan serves as a valuable tool in detecting these kinds of errors, contributing significantly to the development of robust and error-free C++ applications.

Fine-tuning ASan for enhanced control

While ASan is a powerful tool for detecting memory errors in C++ programs, there are scenarios where its behavior needs to be fine-tuned. This fine-tuning is crucial for efficiently managing the analysis process, especially when dealing with complex projects that involve external libraries, legacy code, or specific code patterns.

Suppressing warnings from external libraries

In the context of many projects, the use of external libraries is a common practice. However, these libraries, over which you might not have control, can sometimes contain memory issues. When running tools such as ASan, these issues within the external libraries may get flagged, leading to cluttered diagnostics filled with warnings not directly relevant to your project's code. This can be problematic as it may obscure the real issues within your own code base that need attention.

To mitigate this, ASan offers a useful feature that allows you to suppress warnings specifically coming from these external libraries. This ability to filter out irrelevant warnings is valuable in maintaining a clear focus on fixing issues that are within the scope of your own code base. The implementation of this feature typically involves the use of sanitizer special case lists or specifying certain linker flags during the compilation process. These mechanisms provide a means to tell ASan to ignore certain paths or patterns in the diagnostics, effectively reducing the noise from external sources and aiding in a more targeted and efficient debugging process.

Conditional compilation

There are scenarios in software development where you might want to include specific segments of code only when compiling your program with ASan. This approach can be particularly useful for a variety of purposes, such as incorporating additional diagnostics or modifying memory allocations to make them more compatible or friendly with ASan's operations.

To implement this strategy, you can utilize conditional compilation, a technique that includes or excludes parts of the code based on certain conditions. In the case of ASan, you can check for its presence using the __has_feature macro. This macro evaluates at compile-time whether a particular feature (in this case, ASan) is available in the current compilation context. If ASan is being used, the code within the conditional compilation block will be included in the final executable; otherwise, it will be excluded:

```
#if defined(__has_feature)
#  if __has_feature(address_sanitizer)
// Do something specific for AddressSanitizer
#  endif
#endif
```

This method of conditional compilation allows developers to tailor their code specifically for scenarios where ASan is in use, enhancing the effectiveness of the sanitizer and possibly avoiding issues that might arise only in its presence. It provides a flexible way to adjust the behavior of the program depending on the build configuration, which can be invaluable in complex development environments where different configurations are used for development, testing, and production stages.

Disabling sanitizer for specific lines of code

In the course of developing complex software, there are instances where certain operations might be intentionally performed, even though they could be flagged as errors by ASan. Alternatively, you might have segments of your code base that you wish to exclude from ASan's analysis for specific reasons. This could be due to known benign behaviors in your code that ASan might misinterpret as errors, or parts of the code where the overhead introduced by ASan is not desirable.

To address these scenarios, both GCC and Clang compilers provide a method to selectively disable ASan for particular functions or blocks of code. This is achieved through the use of the `__attribute__ ((no_sanitize("address")))` attribute. By applying this attribute to a function or a specific block of code, you can instruct the compiler to omit ASan instrumentation for that particular segment.

This feature is particularly useful as it allows for granular control over what parts of the code are subject to ASan's scrutiny. It enables developers to fine-tune the balance between thorough error detection and the practical realities of their code's behavior or performance requirements. By judiciously applying this attribute, you can ensure that ASan's analysis is both effective and efficient, focusing its efforts where they are most beneficial.

Utilizing the sanitizer special case list

- **Source files and functions (src and fun)**: ASan allows you to suppress error reports in specified source files or functions. This is particularly useful when you want to ignore certain known issues or third-party code.

- **Globals and types (global and type)**: Additionally, ASan introduces the ability to suppress errors for out-of-bound access to globals with certain names and types. This feature is specifically handy for global variables and class/struct types, allowing more targeted error suppression.

Example of a sanitizer special case list entry

Fine-tuning ASan is an essential aspect of integrating it into a large-scale, complex development environment. It allows developers to customize the behavior of ASan to fit the specific needs of the project, be it by excluding external libraries, conditioning code for ASan builds, or ignoring certain errors to focus on more critical issues. By effectively utilizing these fine-tuning capabilities, teams can harness the full power of ASan to ensure robust and reliable C++ applications. The suppression rules can be set in a textual file as follows:

```
fun:FunctionName   # Suppresses errors from FunctionName
global:GlobalVarName  # Suppresses out-of-bound errors on
```

```
GlobalVarName
type:TypeName  # Suppresses errors for TypeName objects
```

This file can then be passed to the runtime via the `ASAN_OPTIONS` environment variable, such as `ASAN_OPTIONS=suppressions=path/to/suppressionfile`.

Performance overhead of ASan

The use of ASan in detecting memory management issues, such as invalid free operations, is highly beneficial in identifying and resolving potential bugs in C++ applications. However, it's important to be aware of the performance implications of using ASan.

Performance impact, limitations, and recommendations

Integrating ASan into the development and testing process brings with it a certain level of performance overhead. Typically, the slowdown introduced by ASan is in the region of 2x, meaning a program instrumented with ASan may run approximately twice as slowly compared to its non-instrumented version. This increased execution time is primarily due to the additional checks and monitoring that ASan performs to meticulously detect memory errors. Every memory access, along with each memory allocation and deallocation operation, is subject to these checks, inevitably resulting in additional CPU cycles being consumed.

Given this performance impact, ASan is predominantly utilized during the development and testing phases of the software life cycle. This usage pattern represents a trade-off: while there is a performance cost to using ASan, the benefits of catching and fixing critical memory-related errors early in the development process are significant. Early detection of such issues helps in maintaining code quality and can substantially reduce the time and resources required for debugging and fixing bugs later in the cycle.

However, deploying ASan-instrumented binaries in a production environment is generally not recommended, especially in scenarios where performance is a critical factor. The overhead introduced by ASan can impact the application's responsiveness and efficiency. That said, in certain contexts, particularly in applications where reliability and security are of paramount importance, and performance considerations are secondary, using ASan in a production-like environment for thorough testing might be justified. In such cases, the additional assurance of stability and security provided by ASan can outweigh the concerns regarding performance degradation.

ASan is supported on the following:

- Linux i386/x86_64 (tested on Ubuntu 12.04)
- macOS 10.7 – 10.11 (i386/x86_64)
- iOS Simulator
- Android ARM

- NetBSD i386/x86_64

- FreeBSD i386/x86_64 (tested on FreeBSD 11-current)

- Windows 8.1+ (i386/x86_64)

LeakSanitizer (LSan)

LSan is a dedicated memory leak detection tool that is part of the ASan suite but can also be used independently. It is specifically designed to identify memory leaks in C++ programs – situations where allocated memory is not freed, leading to increased memory consumption over time.

Integration with ASan

LSan is often used in conjunction with ASan. When you enable ASan in your build, LSan is automatically enabled as well, providing a comprehensive analysis for both memory errors and leaks.

Standalone mode

If you wish to use LSan without ASan, you can enable it by compiling your program with the -fsanitize=leak flag. This is particularly useful when you want to focus solely on memory leak detection without the overhead of other address sanitizations.

Example of memory leak detection

Consider the following C++ code with a memory leak:

```
int main() {
    int* leaky_memory = new int[100]; // Memory allocated and never
freed
    leaky_memory      = nullptr;      // Memory leaked
    (void)leaky_memory;
    return 0;
}
```

In this example, an array of integers is dynamically allocated and not freed, resulting in a memory leak.

When you compile and run this code with LSan, the output might look something like this:

```
=================================================================
==1743181==ERROR: LeakSanitizer: detected memory leaks

Direct leak of 400 byte(s) in 1 object(s) allocated from:
    #0 0x7fa14b6b6357 in operator new[](unsigned long) ../../../../
src/libsanitizer/asan/asan_new_delete.cpp:102
    #1 0x55888aabd19e in main /home/user/clang-sanitizers/main.cpp:2
```

```
    #2 0x7fa14ae29d8f  (/lib/x86_64-linux-gnu/libc.so.6+0x29d8f)

SUMMARY: AddressSanitizer: 400 byte(s) leaked in 1 allocation(s).
```

This output pinpoints the location and size of the memory leak, aiding in quick and effective debugging.

Platform support

As of the latest information available, LSan is supported on Linux, macOS, and Android. The support can vary based on the toolchain and the version of the compiler being used.

LSan is a valuable tool for C++ developers to identify and resolve memory leaks in their applications. Its ability to be used both in conjunction with ASan and in standalone mode offers flexibility in addressing specific memory-related issues. By integrating LSan into the development and testing process, developers can ensure more efficient memory usage and overall improved application stability.

MemorySanitizer (MSan)

MSan is a dynamic analysis tool, part of the LLVM project, designed to detect the use of uninitialized memory in C++ programs. Uninitialized memory use is a common source of bugs that can lead to unpredictable behavior, security vulnerabilities, and difficult-to-diagnose errors.

To use MSan, compile your program with the -fsanitize=memory flag. This instructs the compiler to instrument the code with checks for uninitialized memory usage. For example:

```
clang++ -fsanitize=memory -g -o your_program your_file.cpp
```

Example code demonstrating uninitialized memory usage

Consider the following simple C++ example:

```cpp
#include <iostream>

int main() {
    int* ptr = new int[10];
    if (ptr[1]) {
        std::cout << "xx\n";
    }
    delete[] ptr;
    return 0;
}
```

In this code, the integers are allocated in the heap but not initialized.

When compiled and run with MSan, the output might look like this:

```
==48607==WARNING: MemorySanitizer: use-of-uninitialized-value
    #0 0x560a37e0f557 in main /home/user/clang-sanitizers/main.cpp:5:9
    #1 0x7fa118029d8f  (/lib/x86_64-linux-gnu/libc.so.6+0x29d8f)
(BuildId: c289da5071a3399de893d2af81d6a30c62646e1e)
    #2 0x7fa118029e3f in __libc_start_main (/lib/x86_64-linux-gnu/
libc.so.6+0x29e3f) (BuildId: c289da5071a3399de893d2af81d6a30c62646e1e)
    #3 0x560a37d87354 in _start (/home/user/clang-sanitizers/build/a.
out+0x1e354) (BuildId: 5a727e2c09217ae0a9d72b8a7ec767ce03f4e6ce)

SUMMARY: MemorySanitizer: use-of-uninitialized-value /home/user/clang-
sanitizers/main.cpp:5:9 in main
```

MSan detects the use of the uninitialized variable and points to the exact location in the code where this occurs.

In this case, a fix can be as simple as initializing the array:

```
int* ptr = new int[10]{};
```

Fine-tuning, performance impact, and limitations

- **Fine-tuning**: MSan's fine-tuning options are similar to those of ASan. Users can refer to the official documentation for detailed customization options.

- **Performance impact**: Typically, using MSan introduces a runtime slowdown of about 3x. This overhead is due to the additional checks that MSan performs to detect uses of uninitialized memory.

- **Supported platforms**: MSan is supported on Linux, NetBSD, and FreeBSD. Its effectiveness in detecting uninitialized memory usage makes it a powerful tool for developers working on these platforms.

- **Limitations**: As with other sanitizers, MSan's runtime overhead makes it most suitable for use in testing environments rather than in production. Additionally, MSan requires that the entire program, including all libraries it uses, be instrumented. This can be a limitation in cases where source code for certain libraries is not available.

MSan is an essential tool for detecting the elusive but potentially critical issue of uninitialized memory usage in C++ programs. By providing detailed reports on where and how such issues occur, MSan enables developers to identify and fix these errors, significantly improving the reliability and security of their applications. Integrating MSan into the development and testing phases, despite its performance impact, is a prudent step toward ensuring robust software quality.

TSan

In the realm of C++ programming, effectively managing concurrency and multithreading is both vital and challenging. Thread-related issues, particularly data races, are notoriously difficult to detect and debug. Unlike other bugs that can often be uncovered through deterministic testing methods such as unit tests, threading issues are elusive and non-deterministic in nature. They may not manifest under every run of a program, leading to unpredictable and erratic behavior that can be extremely hard to replicate and diagnose.

The complexity of thread-related issues

- **Non-deterministic behavior**: Concurrency issues, including data races, deadlocks, and thread leaks, are inherently non-deterministic. This means that they do not consistently reproduce under the same conditions, making them elusive and unpredictable.

- **Challenges in detection**: Traditional testing methods, including comprehensive unit tests, often fail to detect these issues. The outcome of a test involving concurrency can vary from one execution to another, depending on factors such as timing, thread scheduling, and system load.

- **Subtle and severe bugs**: Thread-related bugs can remain dormant, only to surface in production under specific conditions, potentially leading to severe implications such as data corruption, performance degradation, and system crashes.

The necessity of TSan

Given the inherent challenges in managing concurrency in C++, tools such as TSan provided by Clang and GCC become essential. TSan is a sophisticated tool designed to detect threading issues, with a particular focus on data races.

Enabling TSan

- **How to turn TSan on**: To enable TSan, compile your C++ code with the -fsanitize=thread flag. This instructs Clang and GCC to instrument your code for runtime detection of threading issues.

- **Compilation example**:

```
clang++ -fsanitize=thread -g -o your_program your_file.cpp
```

This command will compile your_file.cpp with TSan enabled, ready to detect and report threading issues. Note that it is impossible to turn on both thread and ASans at the same time.

Example of a data race in C++

Consider this simple yet illustrative example:

```cpp
#include <iostream>
#include <thread>

int shared_counter = 0;

void increment_counter() {
    for (int i = 0; i < 10000; ++i) {
        shared_counter++; // Potential data race
    }
}

int main() {
    std::thread t1(increment_counter);
    std::thread t2(increment_counter);
    t1.join();
    t2.join();
    std::cout << "Shared counter: " << shared_counter << std::endl;
    return 0;
}
```

Here, two threads modify the same shared resource without synchronization, leading to a data race.

If we build and run this code with TSan enabled, we get the following output:

```
===================
WARNING: ThreadSanitizer: data race (pid=2560038)
  Read of size 4 at 0x555fd304f154 by thread T2:
    #0 increment_counter() /home/user/clang-sanitizers/main.cpp:8
(a.out+0x13f9)
    #1 void std::__invoke_impl<void, void (*)()>(std::__invoke_other,
void (*&&)()) /usr/include/c++/11/bits/invoke.h:61 (a.out+0x228a)
    #2 std::__invoke_result<void (*)()>::type std::__invoke<void (*)
()>(void (*&&)()) /usr/include/c++/11/bits/invoke.h:96 (a.out+0x21df)
    #3 void std::thread::_Invoker<std::tuple<void (*)()> >::_M_
invoke<0ul>(std::_Index_tuple<0ul>) /usr/include/c++/11/bits/std_
thread.h:259 (a.out+0x2134)
    #4 std::thread::_Invoker<std::tuple<void (*)()> >::operator()() /
usr/include/c++/11/bits/std_thread.h:266 (a.out+0x20d6)
    #5 std::thread::_State_impl<std::thread::_Invoker<std::tuple<void
(*)()> > >::_M_run() /usr/include/c++/11/bits/std_thread.h:211
(a.out+0x2088)
    #6 <null> <null> (libstdc++.so.6+0xdc252)
```

```
  Previous write of size 4 at 0x555fd304f154 by thread T1:
    #0 increment_counter() /home/user/clang-sanitizers/main.cpp:8
(a.out+0x1411)
    #1 void std::__invoke_impl<void, void (*)()>(std::__invoke_other,
void (*&&)()) /usr/include/c++/11/bits/invoke.h:61 (a.out+0x228a)
    #2 std::__invoke_result<void (*)()>::type std::__invoke<void (*)
()>(void (*&&)()) /usr/include/c++/11/bits/invoke.h:96 (a.out+0x21df)
    #3 void std::thread::_Invoker<std::tuple<void (*)()> >::_M_
invoke<0ul>(std::_Index_tuple<0ul>) /usr/include/c++/11/bits/std_
thread.h:259 (a.out+0x2134)
    #4 std::thread::_Invoker<std::tuple<void (*)()> >::operator()() /
usr/include/c++/11/bits/std_thread.h:266 (a.out+0x20d6)
    #5 std::thread::_State_impl<std::thread::_Invoker<std::tuple<void
(*)()> > >::_M_run() /usr/include/c++/11/bits/std_thread.h:211
(a.out+0x2088)
    #6 <null> <null> (libstdc++.so.6+0xdc252)

  Location is global 'shared_counter' of size 4 at 0x555fd304f154
(a.out+0x000000005154)

  Thread T2 (tid=2560041, running) created by main thread at:
    #0 pthread_create ../../../../src/libsanitizer/tsan/tsan_
interceptors_posix.cpp:969 (libtsan.so.0+0x605b8)
    #1 std::thread::_M_start_thread(std::unique_ptr<std::thread::_
State, std::default_delete<std::thread::_State> >, void (*)()) <null>
(libstdc++.so.6+0xdc328)
    #2 main /home/user/clang-sanitizers/main.cpp:14 (a.out+0x1484)

  Thread T1 (tid=2560040, finished) created by main thread at:
    #0 pthread_create ../../../../src/libsanitizer/tsan/tsan_
interceptors_posix.cpp:969 (libtsan.so.0+0x605b8)
    #1 std::thread::_M_start_thread(std::unique_ptr<std::thread::_
State, std::default_delete<std::thread::_State> >, void (*)()) <null>
(libstdc++.so.6+0xdc328)
    #2 main /home/user/clang-sanitizers/main.cpp:13 (a.out+0x146e)

SUMMARY: ThreadSanitizer: data race /home/user/clang-sanitizers/main.
cpp:8 in increment_counter()
==================
Shared counter: 20000
ThreadSanitizer: reported 1 warnings
```

This output from TSan indicates a data race condition in a C++ program. Let's break down the key elements of this report to understand what it's telling us:

- **Type of error**: The report begins with a clear indication that a data race has been detected (WARNING: ThreadSanitizer: data race).

- **Location of accesses:**

  The report specifies two conflicting memory accesses (a read and a previous write) to the same location, both of which constitute the data race.

  The accesses occur at the `0x555fd304f154` memory address, which is identified as a global `shared_counter` variable.

- **Details of conflicting accesses:**

  - **Read access by thread T2:**

    - The read operation is performed by thread T2, as indicated in the report.

    - The exact line of code where this read occurs is pinpointed: `increment_counter()` `/home/user/clang-sanitizers/main.cpp:8`. This means the data race read happens in the `increment_counter` function, specifically at *line 8* of `main.cpp`.

    - The report also provides a stack trace leading up to this read, showing the sequence of function calls.

  - **Write access by thread T1:**

    - Similar to the read access, the report details a write operation by thread T1 to the same global variable.

    - The location of this write is also in the `increment_counter` function at *line 8* of `main.cpp`.

- **Thread creation information:**

  The report includes information about where threads T1 and T2 were created in the program (`main.cpp` at *lines 13* and *14*, respectively). This helps in understanding the program's flow leading to the data race.

- **Summary:**

  The summary restates the nature of the issue: `SUMMARY: ThreadSanitizer: data race /home/user/clang-sanitizers/main.cpp:8 in increment_counter()`.

  This concisely points to the function and file where the data race is detected.

Fine-tuning, performance impact, limitations, and recommendations for TSan

TSan typically introduces a runtime slowdown of approximately 5x-15x. This significant increase in execution time is due to the comprehensive checks performed by TSan to detect data races and other threading issues. Along with the slowdown, TSan also increases memory usage, generally by about 5x-10x. This overhead arises from the additional data structures TSan uses to monitor thread interactions and identify potential race conditions.

This list outlines the limitations and current state of TSan:

- **Beta stage**: TSan is currently in the beta stage. While it has been effective in large C++ programs using pthreads, there is no guarantee of its effectiveness for every scenario.
- **Supported threading models**: TSan supports C++11 threading when compiled with llvm's libc++. This compatibility includes the threading features introduced with the C++11 standard.

TSan is supported by several operating systems and architectures:

- **Android**: aarch64, x86_64
- **Darwin (macOS)**: arm64, x86_64
- FreeBSD
- **Linux**: aarch64, x86_64, powerpc64, powerpc64le
- NetBSD

Support is mainly focused on 64-bit architectures. The support for 32-bit platforms is problematic and not planned.

Fine-tuning TSan

The fine-tuning of TSan is very similar to that of ASan. Users interested in detailed fine-tuning options can refer to the official documentation, which provides comprehensive guidance on customizing TSan's behavior to suit specific needs and scenarios.

Recommendations for using TSan

Due to the overhead in performance and memory, TSan is ideally used during the development and testing phases of a project. Its use in production environments should be carefully evaluated against the performance requirements. TSan is particularly useful in projects with significant multithreaded components, where the likelihood of data races and threading issues is higher. Incorporating TSan into **continuous integration** (**CI**) pipelines can help catch threading issues early in the development cycle, reducing the risk of these bugs making it into production.

TSan is a critical tool for developers dealing with the complexities of concurrency in C++. It provides an invaluable service in detecting elusive threading issues that traditional testing methods often miss. By integrating TSan into the development and testing process, developers can significantly enhance the reliability and stability of their multithreaded C++ applications.

UBSan

UBSan is a dynamic analysis tool designed to detect undefined behavior in C++ programs. Undefined behavior, as defined by the C++ standard, refers to code whose behavior is not prescribed, leading to unpredictable program execution. This can include issues such as integer overflow, division by zero, or misuse of null pointers. Undefined behavior can cause erratic program behavior, crashes, and security vulnerabilities. However, it is often used by compiler developers to optimize code. UBSan is crucial for identifying these problems, which are often subtle and hard to detect through standard testing but can cause significant issues in software reliability and security.

Configuring UBSan

To use UBSan, compile your program with the -fsanitize=undefined flag. This instructs the compiler to instrument the code with checks for various forms of undefined behavior. These commands compile the program with UBSan enabled using either Clang or GCC.

Example code demonstrating undefined behavior

Consider this simple example:

```
#include <iostream>

int main() {
    int x = 0;
    std::cout << 10 / x << std::endl;  // Division by zero, undefined
behavior
    return 0;
}
```

In this code, attempting to divide by zero (10 / x) is an instance of undefined behavior.

When compiled and run with UBSan, the output might include something like this:

```
/home/user/clang-sanitizers/main.cpp:5:21: runtime error: division by
zero
SUMMARY: UndefinedBehaviorSanitizer: undefined-behavior /home/user/
clang-sanitizers/main.cpp:5:21 in
0
```

UBSan detects the division by zero and reports the exact location in the code where this occurs.

Fine-tuning, performance impact, and limitations

- **Fine-tuning**: UBSan provides various options to control its behavior, allowing developers to focus on specific kinds of undefined behavior. Users interested in detailed customization can refer to the official documentation.

- **Performance impact**: The runtime performance impact of UBSan is generally lower compared to tools such as ASan and TSan, but it can vary depending on the types of checks enabled. A typical slowdown is usually minimal.

- **Supported platforms**: UBSan is supported on major platforms such as Linux, macOS, and Windows, making it widely accessible for C++ developers.

- **Limitations**: While UBSan is powerful in detecting undefined behavior, it cannot catch every instance, especially those that are highly dependent on specific program states or hardware configurations.

UBSan is an invaluable tool for C++ developers, aiding in the early detection of subtle yet critical issues that can lead to unstable and insecure software. Its integration into the development and testing process is a proactive step towards ensuring the robustness and reliability of C++ applications. With its minimal performance impact and broad platform support, UBSan is a practical addition to any C++ developer's toolkit.

Dynamic code analysis with Valgrind

Valgrind is a powerful tool for memory debugging, memory leak detection, and profiling. It is instrumental in identifying issues such as memory mismanagement and access errors, which are common in complex C++ programs. Unlike compiler-based tools such as Sanitizers, Valgrind works by running the program in a virtual-machine-like environment, checking for memory-related errors.

Setting up Valgrind

Valgrind can typically be installed from your system's package manager. For example, on Ubuntu, you can install it using `sudo apt-get install valgrind`. To run a program under Valgrind, use the `valgrind ./your_program` command. This command executes your program within the Valgrind environment, where it performs its analysis. No special compilation flags are needed for basic memory checking with Valgrind, but including debugging symbols with `-g` can help make its output more useful.

Memcheck – the comprehensive memory debugger

Memcheck, the core tool of the Valgrind suite, is a sophisticated memory debugger for C++ applications. It combines the functionality of address, memory, and LSans, providing a comprehensive analysis of memory usage. Memcheck detects memory-related errors such as the use of uninitialized memory, improper use of memory allocation and deallocation functions, and memory leaks.

To use Memcheck, no special compilation flags are needed, but compiling with debugging information (using `-g`) can enhance the usefulness of Memcheck's reports. Execute your program with Valgrind by using the `valgrind ./your_program` command. For detecting memory leaks, add `--leak-check=full` for more detailed information. Here is an example command:

```
valgrind --leak-check=full ./your_program
```

Since Memcheck covers a wide range of memory-related issues, I am going to show only an example of detecting a memory leak since they are often the hardest to detect. Let us consider the following C++ code with a memory leak:

```
int main() {
    int* ptr = new int(10); // Memory allocated but not freed
    return 0; // Memory leak occurs here
}
```

Memcheck will detect and report the memory leak, indicating where the memory was allocated and that it was not freed:

```
==12345== Memcheck, a memory error detector
==12345== 4 bytes in 1 blocks are definitely lost in loss record 1 of
1
==12345==    at 0x...: operator new(unsigned long) (vg_replace_
malloc.c:...)
==12345==    by 0x...: main (your_file.cpp:2)
...
==12345== LEAK SUMMARY:
==12345==    definitely lost: 4 bytes in 1 blocks
...
```

Performance impact, fine-tuning, and limitations

It is important to remember that Memcheck can significantly slow down program execution, often by 10-30 times, and increase memory usage. This is due to the extensive checks performed on each memory operation.

Memcheck offers several options to control its behavior. For example, --track-origins=yes can help find the sources of uninitialized memory use, although it may further slow down the analysis.

The main limitation of Memcheck is its performance overhead, which makes it unsuitable for production environments. Additionally, while it is thorough in memory leak detection, it may not catch every instance of uninitialized memory use, especially in complex scenarios or when specific compiler optimizations are applied.

Memcheck stands as a vital tool in the C++ developer's toolkit for memory debugging. By providing a detailed analysis of memory errors and leaks, it plays a critical role in enhancing the reliability and correctness of C++ applications. Despite its performance overhead, Memcheck's benefits in identifying and resolving memory issues make it indispensable for the development and testing phases of software development.

Helgrind – threading error detector

Helgrind is a tool within the Valgrind suite, specifically designed to detect synchronization errors in C++ multithreaded applications. It focuses on identifying race conditions, deadlocks, and misuses of the pthreads API. Helgrind operates by monitoring the interactions between threads, ensuring that shared resources are accessed safely and correctly. Its ability to detect threading errors makes it comparable to TSan but with a different underlying approach and usage.

To use Helgrind, you do not need to recompile your program with special flags (although compiling with -g to include debugging symbols is recommended). Run your program with Valgrind using the --tool=helgrind option. Here is an example command:

```
valgrind --tool=helgrind ./your_program
```

Let us consider the data race example that we analyzed before with TSan:

```
#include <iostream>
#include <thread>

int shared_counter = 0;

void increment_counter() {
    for (int i = 0; i < 10000; ++i) {
        shared_counter++; // Potential data race
    }
}

int main() {
    std::thread t1(increment_counter);
    std::thread t2(increment_counter);
    t1.join();
    t2.join();
    std::cout << "Shared counter: " << shared_counter << std::endl;
    return 0;
}
```

Helgrind will detect and report the data race, showing where the threads are concurrently modifying shared_counter without proper synchronization. In addition to identifying data races, Helgrind's output contains thread creation announcements, stack traces, and other details:

```
valgrind --tool=helgrind ./a.out
==178401== Helgrind, a thread error detector
==178401== Copyright (C) 2007-2017, and GNU GPL'd, by OpenWorks LLP et
al.
```

```
==178401== Using Valgrind-3.18.1 and LibVEX; rerun with -h for
copyright info
==178401== Command: ./a.out
==178401== ---Thread-Announceme
nt----------------------------------------
==178401==
==178401== Thread #3 was created
==178401==    at 0x4CCE9F3: clone (clone.S:76)
==178401==    by 0x4CCF8EE: __clone_internal (clone-internal.c:83)
==178401==    by 0x4C3D6D8: create_thread (pthread_create.c:295)
==178401==    by 0x4C3E1FF: pthread_create@@GLIBC_2.34 (pthread_
create.c:828)
==178401==    by 0x4853767: ??? (in /usr/libexec/valgrind/vgpreload_
helgrind-amd64-linux.so)
==178401==    by 0x4952328: std::thread::_M_start_thread(std::unique_
ptr<std::thread::_State, std::default_delete<std::thread::_State> >,
void (*)()) (in /usr/lib/x86_64-linux-gnu/libstdc++.so.6.0.30)
==178401==    by 0x1093F9: std::thread::thread<void (&)(), ,
void>(void (&)()) (std_thread.h:143)
==178401==    by 0x1092AF: main (main.cpp:14)
==178401==
==178401== ---Thread-Announceme
nt----------------------------------------
==178401==
==178401== Thread #2 was created
==178401== ------------------------------------------------------------
------
==178401==
==178401== Possible data race during read of size 4 at 0x10C0A0 by
thread #3
==178401== Locks held: none
==178401==    at 0x109258: increment_counter() (main.cpp:8)
==178401==    by 0x109866: void std::__invoke_impl<void, void (*)
()>(std::__invoke_other, void (*&&)()) (invoke.h:61)
==178401==    by 0x1097FC: std::__invoke_result<void (*)()>::type
std::__invoke<void (*)()>(void (*&&)()) (invoke.h:96)
==178401==    by 0x1097D4: void std::thread::_Invoker<std::tuple<void
(*)()> >::_M_invoke<0ul>(std::_Index_tuple<0ul>) (std_thread.h:259)
==178401==    by 0x1097A4: std::thread::_Invoker<std::tuple<void (*)
()> >::operator()() (std_thread.h:266)
==178401==    by 0x1096F8: std::thread::_State_impl<std::thread::_
Invoker<std::tuple<void (*)()> > >::_M_run() (std_thread.h:211)
==178401==    by 0x4952252: ??? (in /usr/lib/x86_64-linux-gnu/
libstdc++.so.6.0.30)
==178401==    by 0x485396A: ??? (in /usr/libexec/valgrind/vgpreload_
helgrind-amd64-linux.so)
==178401==    by 0x4C3DAC2: start_thread (pthread_create.c:442)
```

```
==178401==      by 0x4CCEA03: clone (clone.S:100)
==178401==
==178401== This conflicts with a previous write of size 4 by thread #2
==178401== Locks held: none
==178401==      at 0x109261: increment_counter() (main.cpp:8)
==178401==      by 0x109866: void std::__invoke_impl<void, void (*)
()>(std::__invoke_other, void (*&&)()) (invoke.h:61)
==178401==      by 0x1097FC: std::__invoke_result<void (*)()>::type
std::__invoke<void (*)()>(void (*&&)()) (invoke.h:96)
==178401==      by 0x1097D4: void std::thread::_Invoker<std::tuple<void
(*)()> >::_M_invoke<0ul>(std::_Index_tuple<0ul>) (std_thread.h:259)
==178401==      by 0x1097A4: std::thread::_Invoker<std::tuple<void (*)
()> >::operator()() (std_thread.h:266)
==178401==      by 0x1096F8: std::thread::_State_impl<std::thread::_
Invoker<std::tuple<void (*)()> > >::_M_run() (std_thread.h:211)
==178401==      by 0x4952252: ??? (in /usr/lib/x86_64-linux-gnu/
libstdc++.so.6.0.30)
==178401==      by 0x485396A: ??? (in /usr/libexec/valgrind/vgpreload_
helgrind-amd64-linux.so)
==178401==  Address 0x10c0a0 is 0 bytes inside data symbol "shared_
counter"
==178401==
Shared counter: 20000
==178401==
==178401== Use --history-level=approx or =none to gain increased
speed, at
==178401== the cost of reduced accuracy of conflicting-access
information
==178401== For lists of detected and suppressed errors, rerun with: -s
==178401== ERROR SUMMARY: 2 errors from 2 contexts (suppressed: 0 from
0) `
```

Performance impact, fine-tuning, and limitations

Using Helgrind can slow down your program execution significantly (often by 20x or more) due to the detailed analysis of threading interactions. This makes it most suitable for testing environments. Helgrind provides several options to customize its behavior, such as controlling the level of checking or ignoring certain errors. The primary limitation is the performance overhead, making it impractical for use in production. Additionally, Helgrind may produce false positives, especially in complex threading scenarios or when using advanced synchronization primitives not fully understood by Helgrind.

Helgrind is an essential tool for developers working with multithreaded C++ applications, providing insights into challenging concurrency problems. It aids in creating more reliable and thread-safe applications by detecting and helping to resolve complex synchronization issues. While its use may be limited to development and testing phases due to performance overhead, the benefits it offers in enhancing the correctness of multithreaded code are invaluable.

Other notable tools in the Valgrind suite

In addition to Helgrind, the Valgrind suite includes several other tools, each with distinct functionalities catering to different aspects of program analysis and performance profiling.

Data Race Detector (DRD) – a thread error detector

DRD is another tool for detecting thread errors, similar to Helgrind. It focuses specifically on identifying data races in multithreaded programs. While both Helgrind and DRD are designed to detect threading issues, DRD is more optimized for detecting data races and generally has a lower performance overhead compared to Helgrind. DRD might produce fewer false positives in certain scenarios but may not be as thorough as Helgrind in detecting all kinds of synchronization errors.

Cachegrind

Cachegrind is a cache and branch-prediction profiler. It provides detailed information about how your program interacts with the computer's cache hierarchy and the efficiency of branch prediction. This tool is invaluable for optimizing program performance, particularly in CPU-bound applications. It helps identify inefficient memory access patterns and areas of code that can benefit from optimization to improve cache utilization.

Callgrind

Callgrind extends the functionality of Cachegrind by adding call-graph generation capabilities. It records the call history among functions in a program, allowing developers to analyze the execution flow and identify performance bottlenecks. Callgrind is particularly useful for understanding the overall structure and interactions in complex applications.

Massif

Massif is a heap profiler that provides insights into a program's memory usage. It helps developers understand and optimize memory consumption, track down memory leaks, and identify where and how memory allocation occurs within a program.

Dynamic heap analysis tool (DHAT)

The **DHAT** is focused on profiling heap allocation patterns. It's particularly useful for finding inefficient use of heap memory, such as excessive small allocations or short-lived allocations that could be optimized.

Each tool in the Valgrind suite offers unique capabilities for analyzing different aspects of program performance and behavior. From threading issues to memory usage and CPU optimization, these tools provide a comprehensive set of functionalities for enhancing the efficiency, reliability, and correctness of C++ applications. Their integration into the development and testing process allows developers to gain deep insights into their code, leading to well-optimized and robust software solutions.

Summary

Compiler-based sanitizers and Valgrind bring distinct advantages and challenges to the debugging and profiling process.

Compiler-based tools such as ASan, TSan, and UBSan are generally more accessible and easier to integrate into the development workflow. They are "cheaper" in terms of the performance overhead they introduce and are relatively straightforward to configure and use. These sanitizers are integrated directly into the compilation process, making them convenient for developers to employ regularly. Their primary advantage lies in their ability to provide immediate feedback during the development phase, catching errors and issues as the code is being written and tested. However, since these tools perform analysis during runtime, their effectiveness is directly tied to the extent of the test coverage. The more comprehensive the tests, the more effective the dynamic analysis, as only the executed code paths are analyzed. This aspect highlights the importance of thorough testing: the better the test coverage, the more issues these tools can potentially uncover.

Valgrind, on the other hand, offers a more powerful and thorough analysis, capable of detecting a wider range of issues, particularly in memory management and threading. Its suite of tools – Memcheck, Helgrind, DRD, Cachegrind, Callgrind, Massif, and DHAT – provides a comprehensive analysis of various aspects of program performance and behavior. However, this power comes with a cost: Valgrind is generally more complex to use and introduces a significant performance overhead compared to compiler-based tools. The choice of whether to use Valgrind or a compiler-based sanitizer often depends on the specific needs of the project and the issues being targeted. While Valgrind's extensive diagnostics offer deep insights into the program, the ease of use and lower performance cost of compiler-based sanitizers make them more suitable for regular use in a CI pipeline.

In summary, while both compiler-based tools and Valgrind have their place in the dynamic analysis landscape, their differences in diagnostics, ease of use, and performance impact make them suited to different stages and aspects of the software development process. Employing these tools as part of a regular CI pipeline is highly recommended, as it allows for the early detection and resolution of issues, contributing significantly to the overall quality and robustness of the software. The subsequent chapter will delve into tools for measuring test coverage, providing insights into how effectively the code base is being tested and thus complementing the dynamic analysis process.

12
Testing

Software testing stands as a cornerstone in the edifice of software development, holding paramount importance in the assurance of software quality, reliability, and maintainability. It is through the meticulous process of testing that developers can ensure their creations meet the highest standards of functionality and user satisfaction. The inception of any software project is invariably intertwined with the potential for bugs and unforeseen issues; it is testing that illuminates these hidden pitfalls, allowing developers to address them proactively, thereby enhancing the overall integrity and performance of the software.

At the heart of software testing lies a diverse array of methodologies, each tailored to examine distinct facets of the software. Among these, unit testing serves as the foundational layer, focusing on the smallest testable parts of the software to ensure their correct behavior. This granular approach facilitates the early detection of errors, streamlining the development process by enabling immediate corrections. Ascending from the micro to the macro perspective, integration testing takes precedence, wherein the interaction between integrated units is scrutinized. This method is pivotal in identifying issues in the interfacing of components, ensuring seamless communication and functionality within the software.

Progressing further, system testing emerges as a comprehensive examination of the complete and integrated software system. This methodology delves into the software's adherence to specified requirements, offering an overarching assessment of its behavior and performance. It is a crucial phase that validates the software's readiness for deployment, ensuring that it functions correctly in its intended environment. Lastly, acceptance testing marks the culmination of the testing process, where the software is evaluated to determine whether it fulfills the criteria for delivery to end users. This final stage is instrumental in affirming the software's alignment with user needs and expectations, serving as the ultimate testament to its quality and effectiveness.

Embarking on this chapter, you will be guided through the intricate landscape of software testing, gaining insights into the pivotal role it plays in the development life cycle. The exploration will encompass the nuanced distinctions between testing methodologies, shedding light on their unique objectives and the scope of their application. Through this journey, you will acquire a comprehensive understanding of how testing underpins the creation of robust, reliable, and user-centric software, setting the stage for the subsequent chapters that delve deeper into the specifics of unit testing and beyond in the realm of C++.

Test-driven development

Test-driven development, commonly abbreviated as **TDD**, is a modern software development approach that has revolutionized the way code is written and tested. At its core, TDD inverts traditional development methodologies by advocating for the creation of tests before the development of the actual functional code. This paradigm shift is encapsulated in a cyclic process known as "Red-Green-Refactor." Initially, a developer writes a test that defines a desired improvement or a new function, which inevitably fails on the first run – this is the "Red" phase, indicating the absence of the corresponding functionality. Subsequently, in the "Green" phase, the developer crafts the minimum amount of code necessary to pass the test, thereby ensuring that the functionality meets the specified requirements. The cycle culminates in the "Refactor" phase, where the new code is refined and optimized without altering its behavior, thus maintaining the test's successful outcome.

The adoption of TDD brings with it a plethora of advantages that contribute to a more robust and reliable code base. One of the most significant benefits is the marked improvement in code quality. Since TDD necessitates the definition of tests upfront, it inherently encourages a more thoughtful and deliberate design process, reducing the likelihood of bugs and errors. Moreover, tests crafted in the TDD process serve a dual purpose as detailed documentation of the code base. These tests provide clear insights into the code's intended functionality and usage, offering valuable guidance for current and future developers. Additionally, TDD facilitates the design and refactoring of code by ensuring that changes do not inadvertently break existing functionalities, thereby fostering a code base that is both flexible and maintainable.

Despite its numerous benefits, TDD is not without its challenges and potential drawbacks. One of the initial hurdles encountered when adopting TDD is the perceived slowdown in the development process. Writing tests before functionality can feel counterintuitive and may extend the time to deliver features, particularly in the early stages of adoption. Furthermore, TDD demands a steep learning curve, requiring developers to acquire new skills and adapt to a different mindset, which can be a significant investment in time and resources. It's also worth noting that TDD may not be universally applicable or ideal for all scenarios. Certain types of projects, such as those involving complex user interfaces or requiring extensive interaction with external systems, may pose challenges to the TDD methodology, necessitating a more nuanced or hybrid approach to testing.

In conclusion, while TDD presents a transformative approach to software development with its emphasis on test-first methodology, it is essential to weigh its benefits against the potential challenges. The effectiveness of TDD is contingent upon the context of its application, the proficiency of the development team, and the nature of the project at hand. As we delve deeper into the subsequent sections, the nuances of unit testing, integration with testing frameworks, and practical considerations will further illuminate the role of TDD in shaping high-quality, maintainable C++ code bases.

Unit testing in C++

Unit tests are a foundational aspect of TDD in software engineering, playing a pivotal role in the C++ development process. They focus on validating the smallest sections of code, known as units, which are typically individual functions, methods, or classes. By testing these components in isolation, unit tests ensure that each part of the software behaves as intended, which is crucial for the system's overall functionality.

In the TDD framework, unit tests take on an even more significant role. They are often written before the actual code, guiding the development process and ensuring that the software is designed with testability and correctness in mind from the outset. This approach to writing unit tests before the implementation helps in identifying bugs early in the development cycle, allowing for timely corrections that prevent the bugs from becoming more complex or affecting other parts of the system. This proactive bug detection not only saves time and resources but also contributes to the software's stability.

Moreover, unit tests act as a safety net for developers, enabling them to refactor code confidently without fear of breaking existing functionality. This is particularly valuable in TDD, where refactoring is a key step in the cycle of writing a test, making it pass, and then improving the code. Beyond their role in bug detection and facilitating refactoring, unit tests also serve as effective documentation, providing clear insights into the expected behavior of the system. This makes them an invaluable resource for developers, especially those new to the code base. Additionally, the process of writing unit tests in the TDD approach often highlights design improvements, leading to more robust and maintainable code.

C++ unit testing frameworks

The C++ ecosystem is rich with unit testing frameworks designed to facilitate the creation, execution, and maintenance of tests. Among these, Google Test and Google Mock stand out for their comprehensive feature set, ease of use, and integration capabilities with C++ projects. In this section, we'll delve into Google Test and Google Mock, highlighting their key features and syntax, and demonstrate how they can be integrated into a CMake project.

Google Test and Google Mock

Google Test, also known as **GTest**, is Google's framework for writing unit tests in C++. It is widely recognized for its extensive testing capabilities and offers several notable features that facilitate comprehensive and efficient unit testing in C++. Among its key offerings are rich assertions, which enable developers to use a variety of assertions such as EXPECT_EQ and ASSERT_NE to compare expected outcomes with actual results, ensuring precise validation of test conditions. Furthermore, Google Test simplifies the management of common test configurations through test fixtures, which define setup and teardown operations, providing a consistent environment for each test.

Another significant feature is the support for parameterized tests, allowing developers to write a single test and run it with multiple inputs. This approach greatly enhances test coverage without the need for duplicative code. Complementing this, Google Test also supports type-parameterized tests, which permit the execution of the same test logic across different data types, broadening the scope of test coverage even further.

One of the most user-friendly features of Google Test is its automatic test discovery mechanism. This feature eliminates the need for manual test registration, as Google Test automatically identifies and executes tests within the project, streamlining the testing process and saving valuable development time.

Google Mock, also known as **gMock**, complements Google Test by providing a robust mocking framework, which integrates seamlessly to simulate complex object behaviors. This capability is invaluable in creating conditions that mimic real-world scenarios, allowing for more thorough testing of code interactions. With Google Mock, developers gain the flexibility to set expectations on mocked objects, tailoring them to specific needs such as the number of times a function is called, the arguments it receives, and the sequence of calls. This level of control ensures that tests can verify not just the outcomes but also the interactions between different parts of the code.

Furthermore, Google Mock is specifically designed to work in harmony with Google Test, facilitating the creation of comprehensive tests that can leverage both actual objects and their mocked counterparts. This integration simplifies the process of writing tests that are both extensive and reflective of real application behavior, thereby enhancing the reliability and maintainability of the codebase.

Integrating Google Test into a C++ project

We're going to demonstrate how to incorporate Google Test into a CMake project, providing a step-by-step guide to configuring CMake to work with Google Test for unit testing in C++ projects.

To start, ensure that Google Test is included in your project. This can be done by adding Google Test as a submodule in your project's repository or downloading it via CMake. Once Google Test is part of your project, the next step is to configure your CMakeLists.txt file to include Google Test in the build process.

Here's an example of how you might configure your `CMakeLists.txt` file to integrate Google Test via a submodule:

```
git submodule add https://github.com/google/googletest.git external/
googletest
```

Update `CMakeLists.txt` to include Google Test and Google Mock in the build:

```
# Minimum version of CMake
cmake_minimum_required(VERSION 3.14)
project(MyProject)

# GoogleTest requires at least C++14
set(CMAKE_CXX_STANDARD 14)
set(CMAKE_CXX_STANDARD_REQUIRED ON)

# Enable testing capabilities
enable_testing()

# Add GoogleTest to the project
add_subdirectory(external/googletest)

# Include GoogleTest and GoogleMock headers
include_directories(${gtest_SOURCE_DIR}/include ${gmock_SOURCE_DIR}/
include)

# Define your test executable
add_executable(my_tests test1.cpp test2.cpp)

# Link GoogleTest and GoogleMock to your test executable
target_link_libraries(my_tests gtest gtest_main gmock gmock_main)
```

In this configuration, `add_subdirectory(external/googletest)` tells CMake to include Google Test in the build. `include_directories` ensures that the Google Test headers are accessible to your test files. `add_executable` defines a new executable for your tests, and `target_link_libraries` links the Google Test libraries to your test executable.

After configuring `CMakeLists.txt`, you can build and run your tests using CMake and make commands. This setup not only integrates Google Test into your project but also leverages CMake's testing capabilities to automate running the tests.

The following code snippet demonstrates another way to configure CMake to use Google Test, which is by downloading Google Test via CMake's `FetchContent` module. This approach allows CMake to download Google Test during the build process, ensuring that the project's dependencies are automatically managed:

```
cmake_minimum_required(VERSION 3.14)
project(MyProject)

# GoogleTest requires at least C++14
set(CMAKE_CXX_STANDARD 14)
set(CMAKE_CXX_STANDARD_REQUIRED ON)

include(FetchContent)
FetchContent_Declare(
  googletest
  URL https://github.com/google/googletest/
archive/03597a01ee50ed33e9dfd640b249b4be3799d395.zip
)
# For Windows: Prevent overriding the parent project's compiler/linker
settings
set(gtest_force_shared_crt ON CACHE BOOL "" FORCE)
FetchContent_MakeAvailable(googletest)
```

While this example focuses on integrating Google Test with CMake, it's worth noting that Google Test is versatile and can be integrated into other build systems as well, such as Google's own Bazel. For projects using different build systems or for more complex configurations, refer to the official Google Test documentation for comprehensive guidance and best practices. This documentation provides valuable insights into leveraging Google Test across various environments and build systems, ensuring that you can effectively implement unit testing in your C++ projects regardless of the development setup.

Usage of Google Test in C++ projects

Google Test provides a comprehensive suite of functionalities to support various testing needs in C++ development. Understanding how to effectively leverage these features can significantly enhance your testing practices. Let's explore the usage of Google Test through simple examples and explanations.

Writing a simple test

A simple test in Google Test can be written using the `TEST` macro, which defines a test function. Within this function, you can use various assertions to verify the behavior of your code. Here's a basic example:

```
#include <gtest/gtest.h>

int add(int a, int b) {
```

```
        return a + b;
}

TEST(AdditionTest, HandlesPositiveNumbers) {
    EXPECT_EQ(5, add(2, 3));
}
```

In this example, EXPECT_EQ is used to assert that the add function returns the expected sum of two positive numbers. Google Test provides a variety of assertions such as EXPECT_GT (greater than), EXPECT_TRUE (Boolean true), and many others for different testing scenarios.

The key difference between EXPECT_* and ASSERT_* assertions lies in their behavior upon failure. While EXPECT_* assertions allow the test to continue running after a failure, ASSERT_* assertions will halt the current test function immediately upon failure. Use EXPECT_* when subsequent lines of the test do not depend on the success of the current assertion, and ASSERT_* when the failure of an assertion would make the continuation of the test meaningless or potentially cause errors.

Using a test fixture

For tests that require a common setup and teardown for multiple test cases, Google Test offers the concept of a test fixture. This is achieved by defining a class derived from ::testing::Test and then using the TEST_F macro to write tests that use this fixture:

```
class CalculatorTest : public ::testing::Test {
protected:
    void SetUp() override {
        // Code here will be called immediately before each test
        calculator.reset(new Calculator());
    }

    void TearDown() override {
        // Code here will be called immediately after each test
        calculator.reset();
    }

    std::unique_ptr<Calculator> calculator;
};

TEST_F(CalculatorTest, CanAddPositiveNumbers) {
    EXPECT_EQ(5, calculator->add(2, 3));
}

TEST_F(CalculatorTest, CanAddNegativeNumbers) {
```

```
        EXPECT_EQ(-5, calculator->add(-2, -3));
}
```

In this example, `SetUp` and `TearDown` are overridden to provide a common setup (initializing a `Calculator` object) and teardown (cleaning up the `Calculator` object) for each test case. `TEST_F` is used to define test functions that automatically use this setup and teardown, ensuring that each test starts with a fresh `Calculator` instance.

The main function

To run the tests, Google Test requires a main function that initializes the Google Test framework and runs all the tests. Here's an example:

```
#include <gtest/gtest.h>

int main(int argc, char **argv) {
    ::testing::InitGoogleTest(&argc, argv);
    return RUN_ALL_TESTS();
}
```

This main function initializes Google Test, passing the command-line arguments to it, which allows for controlling test execution from the command line. `RUN_ALL_TESTS()` runs all the tests that have been defined and returns 0 if all tests pass or 1 otherwise.

By following these examples and explanations, you can start using Google Test to write comprehensive tests for your C++ projects, ensuring that your code behaves as expected across a wide range of scenarios.

Running Google Test tests

After setting up Google Test with your CMake project and compiling your tests, running them is straightforward. You execute the tests using the `ctest` command in your build directory, which CMake uses to run tests defined in your `CMakeLists.txt` file.

When you run the tests for a `Calculator` class, the standard output to your terminal might look like this if you execute the test binary directly:

```
$ cd path/to/build
[==========] Running 4 tests from 2 test suites.
[----------] Global test environment set-up.
[----------] 2 tests from AdditionTests
[ RUN      ] AdditionTests.HandlesZeroInput
[       OK ] AdditionTests.HandlesZeroInput (0 ms)
[ RUN      ] AdditionTests.HandlesPositiveInput
```

```
[       OK ] AdditionTests.HandlesPositiveInput (0 ms)
[----------] 2 tests from AdditionTests (0 ms total)

[----------] 2 tests from SubtractionTests
[ RUN      ] SubtractionTests.HandlesZeroInput
[       OK ] SubtractionTests.HandlesZeroInput (0 ms)
[ RUN      ] SubtractionTests.HandlesPositiveInput
[       OK ] SubtractionTests.HandlesPositiveInput (0 ms)
[----------] 2 tests from SubtractionTests (0 ms total)

[----------] Global test environment tear-down
[==========] 4 tests from 2 test suites ran. (1 ms total)
[  PASSED  ] 4 tests.
```

This output details each test suite and test case, showing which tests were run ([RUN   ]) and their results ([OK] for passed tests). It provides a clear breakdown of the testing process, including setup and teardown phases, and aggregates the results at the end.

If you run the tests using ctest, the output is more concise by default:

```
$ ctest
Test project /path/to/build
    Start 1: AdditionTests.HandlesZeroInput
1/4 Test #1: AdditionTests.HandlesZeroInput ......    Passed    0.01
sec
    Start 2: AdditionTests.HandlesPositiveInput
2/4 Test #2: AdditionTests.HandlesPositiveInput ...    Passed    0.01
sec
    Start 3: SubtractionTests.HandlesZeroInput
3/4 Test #3: SubtractionTests.HandlesZeroInput .....    Passed    0.01
sec
    Start 4: SubtractionTests.HandlesPositiveInput
4/4 Test #4: SubtractionTests.HandlesPositiveInput ..    Passed    0.01
sec

100% tests passed, 0 tests failed out of 4
```

In this ctest output, each line corresponds to a test case, showing its start order, name, and result. The summary at the end gives a quick overview of the total number of tests, how many passed, and how many failed. This format is useful for getting a quick assessment of your test suite's health without the detailed breakdown provided by the Google Test output.

Advanced features of Google Test

Google Test offers a range of advanced features designed to handle complex testing scenarios, providing developers with powerful tools to ensure their code's robustness. Among these features, one notable capability is the support for *death tests*. Death tests are particularly useful for verifying that your code exhibits the expected behavior when it encounters fatal conditions, such as failed assertions or explicit calls to abort(). This is crucial in scenarios where you want to ensure that your application responds appropriately to unrecoverable errors, enhancing its reliability and safety.

Here's a brief example of a death test:

```
void risky_function(bool trigger) {
    if (trigger) {
        assert(false && "Triggered a fatal error");
    }
}

TEST(RiskyFunctionTest, TriggersAssertOnCondition) {
    EXPECT_DEATH_IF_SUPPORTED(risky_function(true), "Triggered a fatal
error");
}
```

In this example, EXPECT_DEATH_IF_SUPPORTED checks that risky_function(true) indeed causes the program to exit (due to the failed assertion), and it matches the specified error message. This ensures that the function behaves as expected under fatal conditions.

Other advanced features of Google Test include *mocking* for simulating complex object interactions, *parameterized tests* for running the same test logic with various inputs, and *type-parameterized tests* for applying the same test logic across different data types. These features enable comprehensive testing strategies that can cover a wide range of scenarios and inputs, ensuring thorough validation of your code.

For developers seeking to leverage the full potential of Google Test, including its advanced features such as death tests and more, the official Google Test documentation serves as an invaluable resource. It offers detailed explanations, examples, and best practices, guiding you through the nuances of effective test writing and execution in C++ projects. By referring to this documentation, you can deepen your understanding of Google Test's capabilities and integrate them effectively into your testing workflow.

Using gMock in C++ projects

In the world of software testing, particularly within the methodology of TDD, a mock object plays a crucial role. It's designed to mimic the behavior of real objects by implementing the same interface, allowing it to stand in for the actual object in tests. However, the power of a mock object lies in its flexibility; developers can specify its behavior at runtime, including which methods are called, their call order, frequency, argument specifications, and the return values. This level of control turns mock objects into powerful tools for testing interactions and integrations within the code.

Mocks address several challenges in testing complex or interconnected systems. When developing prototypes or tests, relying solely on real objects might not be feasible or practical due to constraints such as external dependencies, execution time, or costs associated with real operations. In such cases, mocks provide a lightweight, controllable substitute that replicates the necessary interactions without the overhead or side effects of the real implementations. They enable developers to focus on the behavior and integration of components rather than their underlying implementations, facilitating more focused and efficient testing.

The distinction between fake objects and mock objects is crucial to understanding their appropriate use cases. While both serve as substitutes for real objects in testing, they have different characteristics and purposes:

- **Fake objects**: These are simplified implementations that mimic real objects but typically take shortcuts for the sake of testing efficiency. An example would be an in-memory database that replicates the functionality of a real database system without persistent storage. Fakes are practical for tests where the exact workings of the real object are not under scrutiny.

- **Mock objects**: Unlike fakes, mocks are pre-programmed with specific expectations that form a contract of how they should be used. They are ideal for testing the interactions between the system under test and its dependencies. For instance, when testing a class that relies on a service, a mock of the service can be used to ensure that the class interacts with the service as expected without actually invoking the service's real implementation.

gMock, Google's framework for creating mock classes in C++, provides a comprehensive solution akin to what jMock and EasyMock offer for Java. With gMock, developers first describe the interface of the object to be mocked using macros, which then generate the mock class implementation. Developers can then instantiate mock objects, setting up their expected behaviors and interactions using gMock's intuitive syntax. During test execution, gMock monitors these mock objects, ensuring that all specified interactions adhere to the defined expectations, and flagging any deviations as errors. This immediate feedback is invaluable for identifying issues in how components interact with their dependencies.

Example of using gMock

In unit testing, particularly when interfacing with network operations, mocking is an invaluable technique. This is exemplified in the case of a `Socket` class, which serves as a foundational element for network communication. The `Socket` class abstracts the functionality of sending and receiving raw byte arrays over a network, providing methods such as `send` and `recv`. Concrete classes such as `TcpSocket`, `UdpSocket`, and `WebSocket` extend this base class to implement specific network protocols. The following code shows the definition of the `Socket` class:

```
class Socket {
public:
    // sends raw byte array of given size, returns number of bytes
sent
    // or -1 in case of error
    virtual ssize_t send(void* data, size_t size) = 0;
    // receives raw byte array of given size, returns number of bytes
received
    // or -1 in case of error
    virtual ssize_t recv(void* data, size_t size) = 0;
};
```

For instance, the `DataSender` class relies on a `Socket` instance to send data. This class is meticulously designed to manage data transmission, attempting retries as necessary and handling various scenarios such as partial data sends, peer-initiated connection closures, and connection errors. The objective in unit testing `DataSender` is to validate its behavior across these different scenarios without engaging in actual network communication. The `DataSender` class is defined as follows:

```
struct DataSentParitally {};
struct ConnectionClosedByPeer {};
struct ConnectionError {};

// Class under test
class DataSender {
    static constexpr size_t RETRY_NUM = 2;
public:
    DataSender(Socket* socket) : _socket{socket} {}

    void send() {
        auto data = std::array<int, 32>{};
        auto bytesSent = 0;
        for (size_t i = 0; i < RETRY_NUM && bytesSent != sizeof(data);
++i) {
            bytesSent = _socket->send(&data, sizeof(data));
            if (bytesSent < 0) {
                throw ConnectionError{};
```

```
            }
            if (bytesSent == 0) {
                throw ConnectionClosedByPeer{};
            }
        }
        if (bytesSent != sizeof(data)) {
            throw DataSentParitally{};
        }
    }

private:
    Socket* _socket;
};
```

This requirement leads us to the use of a MockSocket class, derived from Socket, to simulate network interactions. Here's how MockSocket is defined:

```
class MockSocket : public Socket {
public:
    MOCK_METHOD(ssize_t, send, (void* data, size_t size), (override));
    MOCK_METHOD(ssize_t, recv, (void* data, size_t size), (override));
};
```

The MockSocket class utilizes the MOCK_METHOD macro from gMock to mock the send and recv methods of the Socket class, allowing for the specification of expected behavior during tests. The override keyword ensures that these mock methods correctly override their counterparts in the Socket class.

Setting expectations in gMock is done using constructs such as WillOnce and WillRepeatedly, which define how mock methods behave when invoked:

```
TEST(DataSender, HappyPath) {
    auto socket = MockSocket{};
    EXPECT_CALL(socket, send(_, _)).Times(1).WillOnce(Return(32 *
sizeof(int)));

    auto sender = DataSender(&socket);
    sender.send();
}
```

In this HappyPath test, EXPECT_CALL sets an expectation that send will be called exactly once, successfully transmitting all the data in a single attempt.

```
TEST(DataSender, SendSuccessfullyOnSecondAttempt) {
    auto socket = MockSocket{};
```

```
    EXPECT_CALL(socket, send(_, _)).Times(2)
                            .WillOnce(Return(2 * sizeof(int)))
                            .WillOnce(Return(32 *
sizeof(int)));

    auto sender = DataSender(&socket);
    sender.send();
}
```

This test expects two calls to send: the first transmits only a portion of the data, while the second completes the transmission, simulating a successful send on the second attempt.

The rest of the tests check various error scenarios, such as partial data transmission, connection closure by the peer, and connection errors. Here's an example of a test for the scenario where data is sent partially:

```
TEST(DataSender, DataSentParitally) {
    auto socket = MockSocket{};
    EXPECT_CALL(socket, send(_, _)).Times(2)
                            .WillRepeatedly(Return(2 *
sizeof(int)));

    auto sender = DataSender(&socket);
    EXPECT_THROW(sender.send(), DataSentParitally);
}

TEST(DataSender, ConnectionClosedByPeer) {
    auto socket = MockSocket{};
    EXPECT_CALL(socket, send(_, _)).Times(1)
                            .WillRepeatedly(Return(0 *
sizeof(int)));

    auto sender = DataSender(&socket);
    EXPECT_THROW(sender.send(), ConnectionClosedByPeer);
}

TEST(DataSender, ConnectionError) {
    auto socket = MockSocket{};
    EXPECT_CALL(socket, send(_, _)).Times(1)
                            .WillRepeatedly(Return(-1 *
sizeof(int)));

    auto sender = DataSender(&socket);
```

```
        EXPECT_THROW(sender.send(), ConnectionError);
}
```

Running these tests with gMock and observing the output allows us to confirm the `DataSender` class's behavior under various conditions:

```
[==========] Running 5 tests from 1 test suite.
[----------] Global test environment set-up.
[----------] 5 tests from DataSender
[ RUN      ] DataSender.HappyPath
[       OK ] DataSender.HappyPath (0 ms)
[ RUN      ] DataSender.SendSuccessfullyOnSecondAttempt
[       OK ] DataSender.SendSuccessfullyOnSecondAttempt (0 ms)
[ RUN      ] DataSender.DataSentPartially
[       OK ] DataSender.DataSentPartially (1 ms)
[ RUN      ] DataSender.ConnectionClosedByPeer
[       OK ] DataSender.ConnectionClosedByPeer (0 ms)
[ RUN      ] DataSender.ConnectionError
[       OK ] DataSender.ConnectionError (0 ms)
[----------] 5 tests from DataSender (1 ms total)

[----------] Global test environment tear-down
[==========] 5 tests from 1 test suite ran. (1 ms total)
[  PASSED  ] 5 tests.
```

The output succinctly reports the execution and outcomes of each test, indicating the successful validation of the `DataSender` class's handling of different network communication scenarios. For more comprehensive details on utilizing gMock, including its full suite of features, the official gMock documentation serves as an essential resource, guiding developers through effective mocking strategies in C++ unit testing.

Mocking non-virtual methods via dependency injection

In certain scenarios, you might encounter the need to mock non-virtual methods for unit testing. This can be challenging, as traditional mocking frameworks such as gMock primarily target virtual methods due to C++'s polymorphism requirements. However, one effective strategy to overcome this limitation is through dependency injection, coupled with the use of templates. This approach enhances testability and flexibility by decoupling the class dependencies.

Refactoring for testability

To illustrate this, let's refactor the `Socket` class interface and the `DataSender` class to accommodate the mocking of non-virtual methods. We'll introduce templates to `DataSender` to allow injecting either the real `Socket` class or its mock version.

First, consider a simplified version of the Socket class without virtual methods:

```
class Socket {
public:
    // sends raw byte array of given size, returns number of bytes
sent
    // or -1 in case of error
    ssize_t send(void* data, size_t size);
    // receives raw byte array of given size, returns number of bytes
received
    // or -1 in case of error
    ssize_t recv(void* data, size_t size);
};
```

Next, we modify the DataSender class to accept a template parameter for the socket type, enabling the injection of either a real socket or a mock socket at compile-time:

```
template<typename SocketType>
class DataSender {
    static constexpr size_t RETRY_NUM = 2;

public:
    DataSender(SocketType* socket) : _socket{socket} {}

    void send() {
        auto data = std::array<int, 32>{};
        auto bytesSent = 0;
        for (size_t i = 0; i < RETRY_NUM && bytesSent != sizeof(data);
++i) {
            bytesSent = _socket->send(&data, sizeof(data));
            if (bytesSent < 0) {
                throw ConnectionError{};
            }
            if (bytesSent == 0) {
                throw ConnectionClosedByPeer{};
            }
        }
        if (bytesSent != sizeof(data)) {
            throw DataSentPartially{};
        }
    }

private:
    SocketType* _socket;
};
```

With this template-based design, `DataSender` can now be instantiated with any type that conforms to the `Socket` interface, including mock types.

Mocking with templates

For the mock version of `Socket`, we can define a `MockSocket` class as follows:

```
class MockSocket {
public:
    MOCK_METHOD(ssize_t, send, (void* data, size_t size), ());
    MOCK_METHOD(ssize_t, recv, (void* data, size_t size), ());
};
```

This `MockSocket` class mimics the `Socket` interface but uses gMock's MOCK_METHOD to define mock methods.

Unit testing with dependency injection

When writing tests for `DataSender`, we can now inject `MockSocket` using templates:

```
TEST(DataSender, HappyPath) {
    MockSocket socket;
    EXPECT_CALL(socket, send(_, _)).Times(1).WillOnce(Return(32 *
sizeof(int)));

    DataSender<MockSocket> sender(&socket);
    sender.send();
}
```

In this test, `DataSender` is instantiated with `MockSocket`, allowing the `send` method to be mocked as desired. This demonstrates how templates and dependency injection enable the mocking of non-virtual methods, providing a flexible and powerful approach to unit testing in C++.

This technique, while powerful, requires careful design consideration to ensure that the code remains clean and maintainable. For complex scenarios or further exploration of mocking strategies, the official gMock documentation remains an invaluable resource, offering a wealth of information on advanced mocking techniques and best practices.

Mocking singletons

Despite being considered an anti-pattern due to its potential to introduce a global state and tight coupling in software designs, the Singleton pattern is nevertheless prevalent in many code bases. Its convenience for ensuring a single instance of a class often leads to its use in scenarios such as database connections, where a single, shared resource is logically appropriate.

The Singleton pattern's characteristic of restricting class instantiation and providing a global access point presents a challenge for unit testing, particularly when the need arises to mock the singleton's behavior.

Consider the example of a `Database` class implemented as a singleton:

```cpp
class Database {
public:
    std::vector<std::string> query(uint32_t id) const {
        return {};
    }
    static Database& getInstance() {
        static Database db;
        return db;
    }
private:
    Database() = default;
};
```

In this scenario, the `DataHandler` class interacts with the `Database` singleton to perform operations, such as querying data:

```cpp
class DataHandler {
public:
    DataHandler() {}
    void doSomething() {
        auto& db = Database::getInstance();
        auto result = db.query(42);
    }
};
```

To facilitate testing of the `DataHandler` class without relying on the real `Database` instance, we can introduce a templated variation, `DataHandler1`, that allows injecting a mock database instance:

```cpp
template<typename Db>
class DataHandler1 {
public:
    DataHandler1() {}
    std::vector<std::string> doSomething() {
        auto& db = Db::getInstance();
        auto result = db.query(42);
        return result;
    }
};
```

This approach leverages templates to decouple `DataHandler1` from the concrete `Database` singleton, enabling the substitution of a `MockDatabase` during tests:

```
class MockDatabase {
public:
    std::vector<std::string> query(uint32_t id) const {
        return {"AAA"};
    }
    static MockDatabase& getInstance() {
        static MockDatabase db;
        return db;
    }
};
```

With `MockDatabase` in place, unit tests can now simulate database interactions without hitting the actual database, as demonstrated in the following test case:

```
TEST(DataHandler, check) {
    auto dh = DataHandler1<MockDatabase>{};
    EXPECT_EQ(dh.doSomething(), std::vector<std::string>{"AAA"});
}
```

This test instantiates `DataHandler1` with `MockDatabase`, ensuring that the `doSomething` method interacts with the mock rather than the real database. The expected result is a predefined mock response, making the test predictable and isolated from external dependencies.

This templated solution, a variation of the dependency injection technique discussed earlier, showcases the flexibility and power of templates in C++. It elegantly addresses the challenge of mocking singletons, thereby enhancing the testability of components that depend on singleton instances. For more complex scenarios or further exploration of mocking strategies, referring to the official gMock documentation is advisable, as it offers comprehensive insights into advanced mocking techniques and best practices.

The Nice, the Strict, and the Naggy

In the world of unit testing with gMock, managing the behavior of mock objects and their interactions with the system under test is crucial. gMock introduces three modes to control this behavior: Naggy, Nice, and Strict. These modes determine how gMock handles uninteresting calls – those not matched by any EXPECT_CALL.

Naggy mocks

By default, mock objects in gMock are "naggy." This means that while they warn about uninteresting calls, these calls do not cause the test to fail. The warning serves as a reminder that there might be unexpected interactions with the mock, but it's not critical enough to warrant a test failure. This behavior ensures that tests focus on the intended expectations without being too lenient or too strict about incidental interactions.

Consider the following test scenario:

```
TEST(DataSender, Naggy) {
    auto socket = MockSocket{};
    EXPECT_CALL(socket, send(_, _)).Times(1).WillOnce(Return(32 *
sizeof(int)));

    auto sender = DataSender(&socket);
    sender.send();
}
```

In this case, if there's an uninteresting call to `recv`, gMock issues a warning but the test will pass, marking unanticipated interactions without failing the test.

Nice mocks

`NiceMock` objects go a step further by suppressing warnings for uninteresting calls. This mode is useful when the test's focus is strictly on specific interactions, and other incidental calls to the mock should be ignored without cluttering the test output with warnings.

Using `NiceMock` in a test looks like this:

```
TEST(DataSender, Nice) {
    auto socket = NiceMock<MockSocket>{};
    EXPECT_CALL(socket, send(_, _)).Times(1).WillOnce(Return(32 *
sizeof(int)));

    auto sender = DataSender(&socket);
    sender.send();
}
```

In this `Nice` mode, even if there are uninteresting calls to `recv`, gMock quietly ignores them, keeping the test output clean and focused on the defined expectations.

Strict mocks

On the other end of the spectrum, `StrictMock` objects treat uninteresting calls as errors. This strictness ensures that every interaction with the mock is accounted for by an `EXPECT_CALL`. This mode is particularly useful in tests where precise control over mock interactions is necessary, and any deviation from the expected calls should lead to test failure.

A test using `StrictMock` might look like this:

```
TEST(DataSender, Strict) {
    auto socket = StrictMock<MockSocket>{};
    EXPECT_CALL(socket, send(_, _)).Times(1).WillOnce(Return(32 *
sizeof(int)));

    auto sender = DataSender(&socket);
    sender.send();
}
```

In `Strict` mode, any uninteresting call, such as to `recv`, results in a test failure, enforcing strict adherence to the defined expectations.

Test output and recommended settings

The behavior of these mocking modes is reflected in the test output:

```
Program returned: 1
Program stdout
[==========] Running 3 tests from 1 test suite.
[----------] Global test environment set-up.
[----------] 3 tests from DataSender
[ RUN      ] DataSender.Naggy

GMOCK WARNING:
Uninteresting mock function call - returning default value.
    Function call: recv(0x7ffd4aae23f0, 128)
          Returns: 0
NOTE: You can safely ignore the above warning unless this call should
not happen.  Do not suppress it by blindly adding an EXPECT_CALL() if
you don't mean to enforce the call.  See https://github.com/google/
googletest/blob/master/googlemock/docs/cook_book.md#knowing-when-to-
expect for details.
[        OK ] DataSender.Naggy (0 ms)
[ RUN      ] DataSender.Nice
[        OK ] DataSender.Nice (0 ms)
[ RUN      ] DataSender.Strict
```

```
unknown file: Failure
Uninteresting mock function call - returning default value.
    Function call: recv(0x7ffd4aae23f0, 128)
          Returns: 0
[  FAILED  ] DataSender.Strict (0 ms)
[----------] 3 tests from DataSender (0 ms total)

[----------] Global test environment tear-down
[==========] 3 tests from 1 test suite ran. (0 ms total)
[  PASSED  ] 2 tests.
[  FAILED  ] 1 test, listed below:
[  FAILED  ] DataSender.Strict

 1 FAILED TEST
```

In Naggy mode, the test passes with a warning for uninteresting calls. Nice mode also passes but without any warnings. Strict mode, however, fails the test if there are uninteresting calls.

It is recommended to start with StrickMock and then relax the mode as needed. This approach ensures that tests are initially strict about interactions with mock objects, providing a safety net for unexpected calls. As the test suite matures and the expected interactions become clearer, the mode can be relaxed to Naggy or Nice to reduce noise in the test output.

For further exploration of these modes and advanced mocking techniques, the official gMock documentation provides comprehensive insights and examples, guiding developers through effective mock object management in unit testing.

Throughout this section, we delved into the functionalities and practical applications of Google Test (GTest) and Google Mock (GMock), essential tools for enhancing the testing framework and development workflow of C++ projects. GTest offers a robust environment for creating, managing, and executing unit tests, featuring test fixtures for shared setup and teardown routines, parameterized tests for varied input testing, and type-parameterized tests for applying the same tests across different data types. Its comprehensive assertion library ensures thorough validation of code behavior, contributing to the stability and durability of the software.

Complementing GTest, GMock allows for the seamless creation and utilization of mock objects, enabling isolated component testing by mimicking the behavior of dependencies. This is invaluable in complex systems where direct testing with real dependencies is either impractical or counterproductive. With GMock, developers gain access to a suite of features including automatic mock generation, versatile expectation settings, and detailed behavior verification, enabling in-depth testing of component interactions.

By integrating GTest and GMock into the C++ development life cycle, developers can adopt a robust test-driven approach, ensuring code quality and facilitating continuous testing and integration practices, ultimately leading to more reliable and maintainable software projects.

Other notable C++ unit testing frameworks

Beyond Google Test and Google Mock, the C++ ecosystem is rich with unit testing frameworks, each offering unique features and philosophies. These frameworks cater to various testing needs and preferences, providing developers with multiple options for integrating unit testing into their projects.

Catch2

Catch2 stands out for its simplicity and ease of use, requiring minimal boilerplate code to get started. It adopts a header-only distribution, making it straightforward to integrate into projects. Catch2 supports a variety of testing paradigms, including BDD-style test cases, and offers expressive assertion macros that enhance test readability and intent. Its standout feature is the "Sections" mechanism, which provides a natural way to share setup and teardown code among tests in a flexible and hierarchical manner.

Boost.Test

Part of the extensive Boost libraries, Boost.Test offers robust support for unit testing in C++. It provides a comprehensive assertion framework, test organization facilities, and integration with the Boost build system. Boost.Test can be used in a header-only mode or compiled mode, offering flexibility in its deployment. It's known for its detailed test result reports and wide range of built-in tools for test case management, making it suitable for both small and large-scale projects.

Doctest

Doctest is designed with a focus on simplicity and speed, positioning itself as the lightest feature-rich C++ testing framework. It's particularly appealing for TDD due to its fast compile times. Inspired by Catch2, Doctest offers a similar syntax but aims to be more lightweight and faster to compile, making it ideal for including tests in everyday development without impacting build times significantly.

Google Test versus Catch2 versus Boost.Test versus Doctest

- **Simplicity**: Catch2 and Doctest excel in simplicity and ease of use, with Catch2 offering BDD-style syntax and Doctest being extremely lightweight

- **Integration**: Google Test and Boost.Test will provide more extensive integration capabilities, particularly suited for larger projects with complex testing needs

- **Performance**: Doctest stands out for its compile-time and runtime performance, making it ideal for rapid development cycles

- **Features**: Boost.Test and Google Test come with a more comprehensive set of features out of the box, including advanced test case management and detailed reporting

Choosing the right framework often comes down to project-specific requirements, developer preferences, and the desired balance between simplicity, performance, and feature richness. Developers are encouraged to explore these frameworks further to determine which best fits their unit testing needs, contributing to more reliable, maintainable, and high-quality C++ software.

Good candidates for unit tests

Identifying the optimal candidates for unit testing is pivotal in establishing a robust testing strategy. Unit tests excel when applied to parts of the code base that are well-suited to isolation and fine-grained verification. Here are some key examples and recommendations:

Classes and functions with clear boundaries and well-defined responsibilities are prime candidates for unit testing. These components should ideally embody the Single Responsibility Principle, handling a specific aspect of the application's functionality. Testing these isolated units allows for precise verification of their behavior, ensuring that they perform their intended tasks correctly under various conditions.

Pure functions, which depend solely on their input parameters and produce no side effects, are excellent targets for unit tests. Their deterministic nature – where a given input always results in the same output – makes them straightforward to test and verify. Pure functions are often found in utility libraries, mathematical computations, and data transformation operations.

Components that interact with dependencies through well-defined interfaces are easier to test, especially when those dependencies can be easily mocked or stubbed. This facilitates testing the component in isolation, focusing on its logic rather than the implementation details of its dependencies.

The business logic layer, which encapsulates the core functionality and rules of the application, is typically well-suited for unit testing. This layer often involves calculations, data processing, and decision-making that can be tested in isolation from the user interface and external systems.

While many aspects of an application are suitable for unit testing, it's prudent to recognize scenarios that pose challenges. Components that require complex interactions with external resources, such as databases, filesystems, and network services, might be difficult to effectively mock or might lead to flaky tests due to their reliance on external state or behavior. While mocking can simulate some of these interactions, the complexity and overhead might not always justify the effort in the context of unit testing.

Although unit tests are invaluable for verifying individual components, they have their limitations, especially concerning integrations and end-to-end interactions. For code that is inherently difficult to isolate or requires complex external interactions, **end-to-end (E2E)** tests become crucial. E2E tests simulate real-world usage scenarios, covering the flow from the user interface through to the backend systems and external integrations. In the next section, we will delve into E2E testing, exploring its role in complementing unit tests and providing comprehensive coverage of the application's functionality.

E2E testing in software development

E2E testing is a comprehensive testing approach that evaluates the application's functionality and performance from start to finish. Unlike unit testing, which isolates and tests individual components or units of code, E2E testing examines the application as an integrated whole, simulating real-world user scenarios. This method ensures that all the various components of the application, including its interfaces, databases, networks, and other services, work harmoniously to deliver the desired user experience.

E2E testing frameworks

Given that E2E testing often involves interacting with the application from the outside, it's not confined to the language in which the application is written. For C++ applications, which might be part of a larger ecosystem or serve as backend systems, E2E testing can be conducted using a variety of frameworks across different languages. Some popular E2E testing frameworks include the following:

- **Selenium**: Predominantly used for web applications, Selenium can automate browsers to simulate user interactions with web interfaces, making it a versatile tool for E2E testing

- **Cypress**: Another powerful tool for web applications, Cypress offers a more modern and developer-friendly approach to E2E testing with rich debugging capabilities and a robust API

- **Postman**: For applications exposing RESTful APIs, Postman allows comprehensive API testing, ensuring that the application's endpoints perform as expected under various conditions

When to use E2E testing

E2E testing is particularly valuable in scenarios where the application's components must interact in complex workflows, often involving multiple systems and external dependencies. It's crucial for the following:

- **Testing complex user workflows**: E2E testing shines in validating user journeys that span multiple application components, ensuring a seamless experience from the user's perspective

- **Integration scenarios**: When the application interacts with external systems or services, E2E testing verifies that these integrations work as intended, catching issues that might not be evident in isolation

- **Critical path testing**: For features and pathways that are critical to the application's core functionality, E2E testing ensures reliability and performance under realistic usage conditions

Situations favoring E2E testing

Complex interactions

In situations where the application's components engage in intricate interactions, possibly spanning different technologies and platforms, unit tests might fall short. E2E testing is indispensable for ensuring that the collective behavior of these components aligns with the expected outcomes, especially in:

The architecture outlined in the diagram represents a typical web application with several interconnected services, each serving a distinct role in the system.

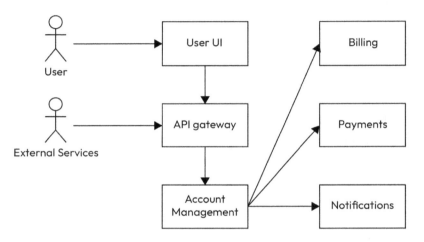

Figure 12.1 – E2E testing

At the frontend, there's a **user UI**, which is the graphical interface where users interact with the application. It's designed to send and receive data to and from the backend services through an **API gateway**. The API gateway acts as an intermediary that routes requests from the user UI to the appropriate backend services and aggregates responses to send back to the UI.

Several backend services are illustrated:

- **Account management**: This service handles user accounts, including authentication, profile management, and other user-related data

- **Billing**: Responsible for managing billing information, subscriptions, and invoicing

- **Payments**: Processes financial transactions, such as credit card processing or interfacing with payment gateways

- **Notifications**: Sends out alerts or messages to users, likely triggered by certain events in the account management or billing services

External services, possibly third-party applications or data providers, can also interact with the API gateway, providing additional functionality or data that supports the main application.

For E2E testing of this system, tests would simulate user actions on the user UI, such as signing up for an account or making a payment. The tests would then verify that the UI correctly sends the appropriate requests through the API gateway to the backend services. Subsequently, the tests would confirm that the user UI responds correctly to the data received from the backend, ensuring that the entire workflow, from the user UI down to notifications, operates as expected. This comprehensive testing approach ensures that each component functions individually and in concert with the rest of the system, delivering a seamless experience for the user.

To summarize, it is essential to consider E2E testing in scenarios where the application's components engage in complex interactions, especially when these interactions span different technologies and platforms. E2E testing ensures that the collective behavior of these components aligns with the expected outcomes, providing a comprehensive assessment of the application's functionality and performance. Here are some of the most common cases when E2E is beneficial:

- **Multi-layered applications**: Applications with multiple layers or tiers, such as client-server architectures, benefit from E2E testing to ensure the layers communicate effectively
- **Distributed systems**: For applications spread across different environments or services, E2E testing can validate the data flow and functionality across these distributed components

Real-world environment testing

One of the primary advantages of E2E testing is its ability to replicate the conditions close to the production environment. This includes testing the application on actual hardware, interacting with real databases, and navigating through the genuine network infrastructure. This level of testing is crucial for the following:

- **Performance validation**: Ensuring that the application performs optimally under expected load conditions and user traffic
- **Security assurance**: Verifying that the application's security measures are effective in a realistic environment, protecting against potential vulnerabilities

E2E testing serves as the final checkpoint before software release, offering a comprehensive assessment of the application's readiness for deployment. By simulating real-world scenarios, E2E testing ensures that the application not only meets its technical specifications but also delivers a reliable and user-friendly experience, making it an essential component of the software development life cycle.

Automatic test coverage tracking tools

In the quest to ensure comprehensive testing of software projects, automatic test coverage tracking tools play a pivotal role. These tools provide invaluable insights into the extent to which the source code of an application is executed during testing, highlighting areas that are well-tested and those that may need additional attention.

Automatic test coverage tracking tools with examples

Ensuring comprehensive test coverage is a cornerstone of reliable software development. Tools such as gcov for the **GNU Compiler Collection (GCC)** and llvm-cov for LLVM projects automate the tracking of test coverage, providing crucial insights into how thoroughly the tests exercise the code.

Tool overview with examples

There are two major tools used for automatic test coverage tracking in C++ projects:

- **gcov**: As an integral part of GCC, gcov analyzes the execution paths taken in your code during test runs. For instance, after compiling a C++ example.cpp file with g++ -fprofile-arcs -ftest-coverage example.cpp, running the corresponding test suite generates coverage data. Running gcov example.cpp afterward produces a report detailing the number of times each line of code was executed.

- **llvm-cov**: Serving a similar purpose within LLVM projects, llvm-cov works with Clang to offer detailed coverage reports. Compiling with clang++ -fprofile-instr-generate -fcoverage-mapping example.cpp and then executing the test binary with LLVM_PROFILE_FILE="example.profraw" ./example prepares the coverage data. llvm-profdata merge -sparse example.profraw -o example.profdata followed by llvm-cov show ./example -instr-profile=example.profdata generates a coverage report for example.cpp.

Integration with C++ projects

Integrating these tools into C++ projects involves compiling the source with coverage flags, executing the tests to generate coverage data, and then analyzing this data to produce reports.

For a project with multiple files, you might compile with the following:

```
g++ -fprofile-arcs -ftest-coverage file1.cpp file2.cpp -o
testExecutable
```

After running ./testExecutable to execute your tests, use gcov file1.cpp file2.cpp to generate coverage reports for each source file.

With `llvm-cov`, the process is similar but tailored for Clang. After compilation and test execution, merging profile data with `llvm-profdata` and generating the report with `llvm-cov` provides a comprehensive view of test coverage.

Interpreting coverage reports

The coverage reports generated by these tools offer several metrics:

- **Line coverage**: This indicates the percentage of code lines executed. For example, a `gcov` report might state `Lines executed:90.00% of 100`, meaning 90 out of 100 lines were run during tests.

- **Branch coverage**: This provides insight into conditional statement testing. A line in the `gcov` report such as `Branches executed:85.00% of 40` shows that 85% of all branches were tested.

- **Function coverage**: This shows the percentage of functions that were called. An entry such as `Functions executed:95.00% of 20` in a `gcov` report indicates that 95% of functions were invoked during testing.

For example, a simplified `gcov` report might look like this:

```
File 'example.cpp'
Lines executed:90.00% of 100
Branches executed:85.00% of 40
Functions executed:95.00% of 20
```

Similarly, an `llvm-cov` report provides detailed coverage metrics, along with the specific lines and branches covered, enhancing the ability to pinpoint areas needing additional tests.

These reports guide developers in improving test coverage by highlighting untested code paths and functions, but they should not be the sole metric for test quality. High coverage with poorly designed tests can give a false sense of security. Effective use of these tools involves not just aiming for high coverage percentages but also ensuring that tests are meaningful and reflective of real-world usage scenarios.

Utilizing hit maps for enhanced test coverage analysis

Hit maps, produced by test coverage tracking tools such as `gcov` and `llvm-cov`, offer a granular view of how tests exercise the code, serving as a detailed guide for developers aiming to improve test coverage. These hit maps go beyond simple percentage metrics, showing precisely which lines of code were executed during tests and how many times, thus enabling a more informed approach to enhance test suites.

Understanding hit maps

A hit map is essentially a detailed annotation of the source code, with each line accompanied by execution counts indicating how many times tests have run that particular line. This level of detail helps identify not only untested parts of the code but also areas that might be over-tested or need more varied testing scenarios.

The .gcov files generated by gcov and the annotated source code produced by llvm-cov provide these hit maps, offering a clear picture of test coverage at the line level.

```
  -:      0:Source:example.cpp
  -:      0:Graph:example.gcno
  -:      0:Data:example.gcda
  -:      0:Runs:3
  -:      0:Programs:1
  3:      1:int main() {
  -:      2:  // Some comment
  2:      3:  bool condition = checkCondition();
  1:      4:  if (condition) {
  1:      5:    performAction();
         ...
```

In this example, line 3 (bool condition = checkCondition();) was executed twice, while the performAction(); line within the if statement was executed once, indicating that the condition was true in one of the test runs.

Similar to gcov, after compiling with clang++ using the -fprofile-instr-generate -fcoverage-mapping flags and executing the tests, llvm-cov can produce a hit map using the llvm-cov show command with the -instr-profile flag pointing to the generated profile data. For example, llvm-cov show ./example -instr-profile=example.profdata example.cpp outputs the annotated source code with execution counts.

The output would resemble the following:

```
example.cpp:
int main() {
    |   3|  // Some comment
    |   2|  bool condition = checkCondition();
    |   1|  if (condition) {
    |   1|    performAction();
    ...
```

Here, the execution count is prefixed to each line, providing a clear picture of test coverage at a glance.

Leveraging hit maps for test improvement

By examining hit maps, developers can identify code sections that are not covered by any test case, indicated by execution counts of zero. These areas represent potential risks for undetected bugs and should be prioritized for additional testing. Conversely, lines with exceptionally high execution counts might indicate areas where tests are redundant or overly focused, suggesting an opportunity to diversify test scenarios or refocus testing efforts on less-covered parts of the code base.

Incorporating hit map analysis into regular development workflows encourages a proactive approach to maintaining and enhancing test coverage, ensuring that tests remain effective and aligned with the evolving code base. As with all testing strategies, the goal is not merely to achieve high coverage numbers but to ensure that the test suite comprehensively validates the software's functionality and reliability in a variety of scenarios.

Incorporating hit maps into the development workflow has been made even more accessible with the advent of **integrated development environment** (**IDE**) plugins that integrate coverage visualization directly into the coding environment. A notable example is the "Code Coverage" plugin by Markis Taylor for **Visual Studio Code** (**VSCode**). This plugin overlays hit maps onto the source code within the VSCode editor, providing immediate, visual feedback on test coverage.

The "Code Coverage" plugin processes coverage reports generated by tools such as `gcov` or `llvm-cov` and visually annotates the source code in VSCode. Lines of code covered by tests are highlighted, typically in green, while uncovered lines are marked in red. This immediate visual representation allows developers to quickly identify untested code regions without leaving the editor or navigating through external coverage reports.

Recommendations for code coverage

Code coverage is a vital metric in the realm of software testing, providing insights into the extent to which the code base is exercised by the test suite. For C++ projects, leveraging tools such as `gcov` for GCC and `llvm-cov` for LLVM projects can offer detailed coverage analysis. These tools are adept at not only tracking coverage from unit tests but also from E2E tests, allowing for a comprehensive assessment of test coverage across different testing levels.

A robust testing strategy involves a combination of focused unit tests, which validate individual components in isolation, and broader E2E tests, which assess the system's functionality as a whole. By employing `gcov` or `llvm-cov`, teams can aggregate coverage data from both testing types, providing a holistic view of the project's test coverage. This combined approach helps identify areas of the code that are either under-tested or not tested at all, guiding efforts to enhance the test suite's effectiveness.

It is recommended to keep a vigilant eye on code coverage metrics and strive to prevent any decrease in coverage percentages. A decline in coverage might indicate new code being added without adequate testing, potentially introducing undetected bugs into the system. To mitigate this risk, teams should integrate coverage checks into their **continuous integration** (**CI**) pipelines, ensuring that any changes that reduce coverage are promptly identified and addressed.

Periodically, it's beneficial to allocate time specifically for increasing test coverage, especially in areas identified as critical or risky. This might involve writing additional tests for complex logic, edge cases, or error-handling paths that were previously overlooked. Investing in coverage improvement initiatives not only enhances the software's reliability but also contributes to a more maintainable and robust code base in the long term.

Summary

This chapter provided a thorough overview of testing in C++, covering essential topics from unit testing basics to advanced E2E testing. You learned about unit testing's role in ensuring individual components work correctly and how tools such as Google Test and Google Mock help write and manage these tests effectively. The chapter also touched on mocking techniques for simulating complex behaviors in tests.

Additionally, the importance of tracking test coverage using tools such as `gcov` and `llvm-cov` was discussed, emphasizing the need to maintain and improve coverage over time. E2E testing was highlighted as crucial for checking the entire application's functionality, complementing the more focused unit tests.

By exploring different C++ testing frameworks, the chapter offered insights into the various tools available for developers, helping them choose the right ones for their projects. In essence, this chapter equipped you with the knowledge to implement comprehensive and effective testing strategies in your C++ development endeavors, contributing to the creation of reliable and robust software.

In the next chapter, we will explore modern approaches to third-party management in C++, including Docker-based solutions and available package managers.

13
Modern Approach to Managing Third Parties

In modern software development, the reliance on third-party libraries is virtually inescapable. From foundational components such as OpenSSL for secure communication, and Boost for extensive C++ libraries, to even the standard library that forms the bedrock of C++ programming, external libraries are integral to building functional and efficient applications. This dependency underscores the importance of understanding how third-party libraries are managed within the C++ ecosystem.

Given the complexity and diversity of these libraries, it's crucial for developers to grasp the basics of third-party library management. This knowledge not only aids in the seamless integration of these libraries into projects but also influences deployment strategies. The compilation method of a library, whether static or dynamic, directly impacts the number of files deployed and the overall footprint of the application.

Unlike some other programming languages that benefit from a standardized library ecosystem, C++ presents a unique challenge due to the absence of such a unified system. This chapter delves into the existing solutions for third-party library management in C++, exploring tools such as vcpkg, Conan, and others. By examining these tools, we aim to provide insights into which solution might best fit your project's needs, considering factors such as platform compatibility, ease of use, and the scope of the library catalog.

As we navigate through these solutions, our goal is to equip you with the knowledge to make informed decisions about integrating and managing third-party libraries in your C++ projects, thereby enhancing your development workflow and the quality of your software.

302 Modern Approach to Managing Third Parties

Overview of linking and shared V threads::ThreadsS static libraries

In the context of C and C++ development, third-party entities are external libraries or frameworks that developers integrate into their projects. These entities serve to improve functionality or utilize existing solutions. These third-party components can vary significantly in scope, from minimal utility libraries to comprehensive frameworks offering a broad range of features.

The process of integrating third-party libraries into a project involves using header files that outline the interfaces of these libraries. These header files contain the declarations of classes, functions, and variables provided by the library, allowing the compiler to understand the required signatures and structures for successful compilation. Including a header file in a C++ source file essentially concatenates the contents of the header file to the point of inclusion, enabling access to the library's interfaces without embedding the actual implementation within the source file.

The implementation of these libraries is supplied through compiled object code, typically distributed as either static libraries or shared libraries. Static libraries are archives of object files that are directly incorporated into the final executable by the linker, resulting in a larger executable size due to the embedding of the library code. On the other hand, shared libraries, known as **Dynamic Link Libraries (DLLs)** on Windows or **Shared Objects (SOs)** on Unix-like systems, are not embedded into the executable. Instead, references to these libraries are included, and the operating system loads them into memory at runtime. This mechanism allows multiple applications to utilize the same library code, conserving memory.

Shared libraries were designed to facilitate the sharing of common libraries, such as libc or C++ standard libraries, among multiple applications. This practice is especially advantageous for frequently utilized libraries. This design also theoretically allows users to update shared libraries without needing to upgrade the entire application. However, in practice, this is not always seamless and can introduce compatibility issues, making it less advantageous for applications to provide their dependencies as shared libraries. Furthermore, opting for shared libraries over static ones can reduce linker time, as the linker does not need to embed the library code into the executable, which can speed up the build process.

The linker plays a pivotal role in this process, merging various object files and libraries into a single executable or library, and resolving symbol references along the way to ensure the final binary is complete and executable.

The choice between static and dynamic linking significantly affects application performance, size, and deployment strategies. Static linking simplifies deployment by creating self-contained executables but at the cost of larger file sizes and the necessity to recompile for library updates. Dynamic linking, while reducing memory usage by sharing library code among applications and facilitating easier library updates, introduces complexities in deployment to ensure all dependencies are met.

Given the complexities associated with linking external shared objects and the widespread use of templated code in C++, many library developers have started to prefer supplying their libraries as "header-only" libraries. A header-only library is a library that is entirely contained within header files, with no separate implementation files or precompiled binaries. This means that all the code, including function and class definitions, is included in the header files.

This approach simplifies the integration process significantly. When a developer includes a header file from a header-only library, they are not just including interface declarations but the entire implementation. Consequently, there is no need for separate compilation or linking of the library's implementation; the compiler includes and compiles the library's code directly into the developer's source code when the header file is included. This direct inclusion can lead to more efficient inlining and optimizations by the compiler, potentially resulting in faster executable code due to the elimination of function call overheads.

However, it's worth noting that while header-only libraries offer convenience and ease of integration, they also have some downsides. Since the entire library is included and compiled with each source file that includes it, this can lead to increased compilation times, especially for large libraries or projects that include the library in multiple files. Furthermore, any change in the header file necessitates recompiling all source files that include it, which can further increase development time.

Despite its drawbacks, the header-only approach in C++ is highly attractive to many developers and users due to its simplicity of distribution and use. Additionally, it helps avoid linking issues and offers benefits for template-heavy libraries. This model is especially prevalent in libraries where heavy use of templates is made, such as those providing metaprogramming facilities, because templates must be available in their entirety to the compiler at compile time, making the header-only model a natural fit.

In essence, the management of third-party dependencies in C++ projects involves a deep understanding of header files, static and shared libraries, and the intricacies of the linking process. Developers must carefully consider the trade-offs between static and dynamic linking in the context of application requirements and deployment environments, balancing factors such as performance, size, and ease of maintenance.

Managing third-party libraries in C++

Managing third-party libraries is a critical aspect of C++ development. While there is no standardized package manager for C++, various methods and tools have been adopted to streamline this process, each with its own set of practices and supported platforms.

Installing libraries with OS package managers

Many developers rely on the operating system's package manager to install third-party libraries. On Ubuntu and other Debian-based systems, apt is commonly used:

```
sudo apt install libboost-all-dev
```

For Red Hat-based systems, yum or its successor dnf is the go-to option:

```
sudo yum install boost-devel
```

On macOS, Homebrew is a popular choice for managing packages:

```
brew install boost
```

Windows users often turn to Chocolatey or vcpkg (the latter also functions as a general C++ library manager beyond just Windows):

```
choco install boost
```

These OS package managers are convenient for common libraries but might not always offer the latest version or specific configurations needed for development.

Using Git as a third-party manager via submodules

Git submodules allow developers to include and manage the source code of third-party libraries directly within their repositories. This method is advantageous for ensuring all team members and the build system use an exact version of a library. A typical workflow for adding a submodule and integrating it with CMake might look like this:

```
git submodule add https://github.com/google/googletest.git external/
googletest
git submodule update --init
```

In CMakeLists.txt, you'd include the submodule:

```
add_subdirectory(external/googletest)
include_directories(${gtest_SOURCE_DIR}/include ${gtest_SOURCE_DIR})
```

This method tightly couples your project with specific versions of the library and facilitates tracking updates through Git.

Using CMake FetchContent to download libraries

CMake's FetchContent module provides a more flexible alternative to submodules by downloading dependencies at configure time, without the need to include them directly in your repository:

```
include(FetchContent)
FetchContent_Declare(
  json
  GIT_REPOSITORY https://github.com/nlohmann/json.git
  GIT_TAG v3.7.3
```

```
)
FetchContent_MakeAvailable(json)
```

This approach differs from Git submodules by not requiring the library's source code to be present in your repository or updating it manually. `FetchContent` dynamically retrieves the specified version, making it easier to manage and update dependencies.

Conan – advanced dependency management

Conan is a powerful package manager for C and C++ that simplifies the process of integrating third-party libraries and managing dependencies across various platforms and configurations. It stands out for its ability to handle multiple versions of libraries, complex dependency graphs, and different build configurations, making it an essential tool for modern C++ development.

Conan configuration and features

Conan's configuration is stored in `conanfile.txt` or `conanfile.py`, where developers specify the required libraries, versions, settings, and options. This file serves as the manifest for project dependencies, enabling precise control over the libraries used in a project.

Key features:

- **Multi-platform support**: Conan is designed to work on Windows, Linux, macOS, and FreeBSD, offering a consistent experience across different operating systems

- **Build configuration management**: Developers can specify settings such as compiler version, architecture, and build type (debug, release) to ensure compatibility and optimal builds for their projects

- **Version handling**: Conan can manage multiple versions of the same library, allowing projects to depend on specific versions as needed

- **Dependency resolution**: It automatically resolves and downloads transitive dependencies, ensuring that all required libraries are available for the build process

Library locations and Conan Center

The primary repository for Conan packages is **Conan Center**, an extensive collection of open source C and C++ libraries. Conan Center is the go-to place to find and download packages, but developers can also specify custom or private repositories for their projects.

In addition to Conan Center, companies and development teams can host their own Conan servers or use services such as Artifactory to manage private or proprietary packages, enabling a centralized approach to dependency management within an organization.

Configuring static or dynamic linking

Conan allows developers to specify whether to use static or dynamic linking for libraries. This is typically done through options in `conanfile.txt` or `conanfile.py`. Here's an example:

```
[options]
Poco:shared=True  # Use dynamic linking for Poco
Or in conanfile.py:
class MyProject(ConanFile):
    requires = "poco/1.10.1"
    default_options = {"poco:shared": True}
```

These settings instruct Conan to download and use the dynamic version of the specified libraries. Similarly, setting the option to `False` would favor static libraries. It's essential to note that not all packages will support both linking options, depending on how they were packaged for Conan.

Extending Conan with custom packages

One of the strengths of Conan is its extensible nature. If a required library is not available in Conan Center or does not meet specific needs, developers can create and contribute their own packages. Conan provides a Python-based development kit for creating packages, which includes tools for defining build processes, dependencies, and package metadata.

To create a Conan package, developers define `conanfile.py` that describes how to source, build, and package the library. This file includes methods such as `source()`, `build()`, and `package()` that Conan executes during the package creation process.

Once a package is developed, it can be shared through Conan Center by submitting it for inclusion, or it can be distributed through private repositories to maintain control over distribution and usage.

Conan's flexibility, support for multiple platforms and configurations, and its comprehensive package repository make it an invaluable tool for C++ developers. By leveraging Conan, teams can streamline their dependency management process, ensuring consistent, reproducible builds across different environments. The ability to configure static or dynamic linking, coupled with the option to extend the repository with custom packages, underscores Conan's adaptability to diverse project requirements. Whether working with widely-used open source libraries or specialized proprietary code, Conan provides a robust framework for managing C++ dependencies efficiently and effectively.

Conan is a dedicated C++ package manager that excels in managing different versions of libraries and their dependencies. It operates independently of the operating system's package manager and provides a high level of control and flexibility. A typical Conan workflow involves creating `conanfile.txt` or `conanfile.py` to declare dependencies.

CMake integration

CMake is widely used in C++ projects for its powerful scripting capabilities and cross-platform support. Integrating Conan with CMake can significantly streamline the process of managing dependencies. Here's how you can achieve this integration:

- **Conan CMake wrapper**: Conan provides a CMake wrapper script that automates the integration. To use it, include the conanbuildinfo.cmake file generated by Conan in your project's CMakeLists.txt:

```
include(${CMAKE_BINARY_DIR}/conanbuildinfo.cmake)
conan_basic_setup(TARGETS)
```

This script sets up the necessary include paths and library paths, and defines the dependencies managed by Conan, making them available to your CMake project.

- **Using targets**: The TARGETS option in conan_basic_setup() generates CMake targets for your Conan dependencies, allowing you to link against them using the target_link_libraries() function in CMake:

```
target_link_libraries(my_project_target CONAN_PKG::poco)
```

This approach provides a clean and explicit way to link your project's targets against the libraries managed by Conan.

Other build system integration

Conan's flexibility extends to other build systems as well, making it adaptable to various project requirements:

- **Makefiles**: For projects using Makefiles, Conan can generate the appropriate variables for include paths, library paths, and flags that can be included in a Makefile:

```
include conanbuildinfo.mak
```

- **MSBuild**: For projects using MSBuild (common in Windows environments), Conan can generate .props files that can be imported into Visual Studio projects, providing a seamless integration with the MSBuild ecosystem.

- **Bazel, Meson, and others**: While direct support for some build systems such as Bazel or Meson might require custom integration scripts or tools, the Conan community often contributes generators and tools to bridge these gaps, extending Conan's reach to virtually any build system.

Custom integration

For build systems without direct support or for projects with unique requirements, Conan offers the ability to customize the generated files or even write custom generators. This allows developers to tailor the integration to their specific build process, making Conan a highly adaptable tool for dependency management.

Conclusion

The integration of Conan with CMake and other build systems underscores its versatility as a package manager for C++ projects. By providing straightforward mechanisms to incorporate dependencies into various build environments, Conan not only simplifies dependency management but also enhances build reproducibility and consistency across different platforms and configurations. Whether you're working with a widely used build system such as CMake or a more specialized setup, Conan's flexible integration options ensure that you can maintain an efficient and streamlined development workflow.

vcpkg

vcpkg, developed by Microsoft, is a cross-platform C++ package manager that simplifies the process of acquiring and building C++ open source libraries. It is designed to work seamlessly with CMake and other build systems, providing a straightforward and consistent way to manage C++ library dependencies.

Key differences from Conan

While both vcpkg and Conan are aimed at simplifying dependency management in C++ projects, there are notable differences in their approach and ecosystem:

- **Origin and backing**: vcpkg was created and is maintained by Microsoft, which ensures tight integration with Visual Studio and the MSBuild system, although it remains fully functional and useful across different platforms and development environments.

- **Package sources**: vcpkg focuses on compiling from source, ensuring that libraries are built with the same compiler and settings as the consuming project. This approach contrasts with Conan, which can manage precompiled binaries, allowing for quicker integration but potentially leading to binary incompatibility issues.

- **Integration**: vcpkg integrates natively with CMake and Visual Studio, providing manifest files for project-level integration. This can make it particularly attractive for projects already using these tools, offering a more seamless integration experience.

- **Ecosystem and libraries**: Both package managers boast a large collection of available libraries, but their ecosystems might differ slightly due to the community and backing of each project.

Operating system support

vcpkg is designed to be cross-platform, with support for the following:

- Windows
- Linux
- macOS

This wide range of support makes it a versatile option for developers working in diverse development environments.

Example of configuring a project with vcpkg

To illustrate the use of vcpkg in a project, let's go through a simple example of integrating a library, such as the JSON for Modern C++ library (nlohmann-json), into a C++ project using CMake.

Clone the vcpkg repository and run the bootstrap script:

```
git clone https://github.com/Microsoft/vcpkg.git
cd vcpkg
./bootstrap-vcpkg.sh  # Use bootstrap-vcpkg.bat on Windows
Install nlohmann-json using vcpkg:
./vcpkg install nlohmann-json
```

vcpkg will download and compile the library, making it available for projects.

To use vcpkg with a CMake project, you can set the CMAKE_TOOLCHAIN_FILE variable to the path of the vcpkg.cmake toolchain file when configuring your project:

```
cmake -B build -S . -DCMAKE_TOOLCHAIN_FILE=[vcpkg root]/scripts/
buildsystems/vcpkg.cmake
Replace [vcpkg root] with the path to your vcpkg installation.
In your CMakeLists.txt, find and link against the nlohmann-json
package:
cmake_minimum_required(VERSION 3.0)
project(MyVcpkgProject)

find_package(nlohmann_json CONFIG REQUIRED)

add_executable(my_app main.cpp)
target_link_libraries(my_app PRIVATE nlohmann_json::nlohmann_json)
In your main.cpp, you can now use the nlohmann-json library:
#include <nlohmann/json.hpp>
```

```
int main() {
    nlohmann::json j;
    j["message"] = "Hello, world!";
    std::cout << j << std::endl;
    return 0;
}
```

vcpkg, with its emphasis on source-based distribution and integration with CMake and Visual Studio, offers a robust solution for C++ developers looking to manage library dependencies effectively. Its simplicity, coupled with the backing of Microsoft, makes it a compelling choice for projects that prioritize consistency with the build environment and seamless integration with existing Microsoft tools. While it shares common goals with Conan in simplifying dependency management, the choice between vcpkg and Conan may come down to specific project requirements, preferred workflow, and the development ecosystem.

Utilizing Docker for C++ builds

A notable shortfall within C++ is its absence of an inherent mechanism for managing dependencies. Consequently, the incorporation of third-party elements is achieved through a heterogeneous array of methodologies: the utilization of package managers provided by Linux distributions (for instance, apt-get), the direct installation via make install, the inclusion of third-party libraries as Git submodules and their subsequent compilation within the project's source tree, or the adoption of package management solutions such as Conan or Vcpkgvcpkg.

Regrettably, each of these methods comes with its own set of drawbacks:

- The installation of dependencies directly on a development machine tends to compromise the cleanliness of the environment, rendering it dissimilar to those of CI/CD pipelines or production environments – a discrepancy that becomes more pronounced with each update of third-party components.

- It is often a formidable task to ensure uniformity in the versions of compilers, debuggers, and other tools utilized by all developers. This lack of standardization can culminate in a scenario where a build executes successfully on an individual developer's machine yet fails within the CI/CD environment.

- The practice of integrating third-party libraries as Git submodules and compiling them within the project's source directory poses a challenge, particularly when dealing with substantial libraries (such as Boost, Protobuf, Thrift, etc.). This method can lead to a significant deceleration of the build process, to the extent that developers may hesitate to clear the build directory or to alternate between branches.

- Package management solutions such as Conan may not always offer the desired version of a specific dependency, and the inclusion of such a version necessitates the authoring of additional code in Python, which, in my opinion, is unduly burdensome.

A single isolated and reproducible build environment

The optimal resolution for the aforementioned challenges involves the formulation of a Docker image, embedded with all requisite dependencies and tools, such as compilers and debuggers, to facilitate the project's compilation within a container derived from this image.

This particular image serves as the cornerstone for a **singular** build environment that is uniformly employed by developers on their respective workstations as well as on CI/CD servers, effectively eliminating the all-too-common discrepancy of "it works on my machine but fails at CI!".

Owing to the encapsulated nature of the build process within the container, it remains impervious to any external variables, tools, or configurations peculiar to an individual developer's local setup, thereby rendering the build environment **isolated**.

In an ideal scenario, Docker images are meticulously labeled with meaningful version identifiers, enabling users to seamlessly transition between different environments by retrieving the appropriate image from the registry. Furthermore, in the event that an image is no longer available in the registry, it's worth noting that Docker images are constructed from Dockerfiles, which are typically maintained within Git repositories. This ensures that, should the need arise, there is always the feasibility to reconstruct the image from a previous version of the Dockerfile. This attribute of the Dockerized build framework lends it a characteristic of being **reproducible**.

Creating the build image

We will embark on developing a straightforward application and compile it within a container. The essence of the application is to display its size utilizing `boost::filesystem`. The selection of Boost for this demonstration is intentional, aiming to illustrate the integration of Docker with a "heavy" third-party library:

```cpp
#include <boost/filesystem/operations.hpp>
#include <iostream>

int main(int argc, char *argv[]) {
    std::cout << "The path to the binary is: "
              << boost::filesystem::absolute(argv[0])
              << ", the size is:" << boost::filesystem::file_
size(argv[0]) << '\n';
    return 0;
}
```

The CMake file is quite simple:

```cmake
cmake_minimum_required(VERSION 3.10.2)

project(a.out)
```

```
set(CMAKE_CXX_STANDARD 17)
set(CMAKE_CXX_STANDARD_REQUIRED ON)

# Remove for compiler-specific features
set(CMAKE_CXX_EXTENSIONS OFF)

string(APPEND CMAKE_CXX_FLAGS " -Wall")
string(APPEND CMAKE_CXX_FLAGS " -Wbuiltin-macro-redefined")
string(APPEND CMAKE_CXX_FLAGS " -pedantic")
string(APPEND CMAKE_CXX_FLAGS " -Werror")

# clangd completion
set(CMAKE_EXPORT_COMPILE_COMMANDS ON)

include_directories(${CMAKE_SOURCE_DIR})
file(GLOB SOURCES "${CMAKE_SOURCE_DIR}/*.cpp")

add_executable(${PROJECT_NAME} ${SOURCES})

set(Boost_USE_STATIC_LIBS        ON) # only find static libs
set(Boost_USE_MULTITHREADED      ON)
set(Boost_USE_STATIC_RUNTIME     OFF) # do not look for boost libraries
linked against static C++ std lib

find_package(Boost REQUIRED COMPONENTS filesystem)

target_link_libraries(${PROJECT_NAME}
    Boost::filesystem
)
```

> **Note**
>
> In this example, Boost is linked statically since it is required if the target machine does not have the right version of Boost pre-installed; this recommendation applies to all dependencies pre-installed in the Docker image.

The Dockerfile employed for this task is notably uncomplicated:

```
FROM ubuntu:18.04
LABEL Description="Build environment"

ENV HOME /root

SHELL ["/bin/bash", "-c"]
```

```
RUN apt-get update && apt-get -y --no-install-recommends install \
    build-essential \
    clang \
    cmake \
    gdb \
    wget

# Let us add some heavy dependency
RUN cd ${HOME} && \
    wget --no-check-certificate --quiet \
        https://boostorg.jfrog.io/artifactory/main/release/1.77.0/
source/boost_1_77_0.tar.gz && \
        tar xzf ./boost_1_77_0.tar.gz && \
        cd ./boost_1_77_0 && \
        ./bootstrap.sh && \
        ./b2 install && \
        cd .. && \
         rm -rf ./boost_1_77_0
```

To ensure that its name is distinctive and does not overlap with existing Dockerfiles, while also clearly conveying its purpose, I have named it DockerfileBuildEnv:

```
$ docker build -t example/example_build:0.1 -f DockerfileBuildEnv .
Here is supposed to be a long output of boost build
```

*Note that the version is not the "latest" but has a meaningful name (e.g., 0.1).

Once the image has been successfully constructed, we are positioned to proceed with the project's build process. The initial step involves initiating a Docker container that is based on our crafted image, followed by the execution of the Bash shell within this container:

```
$ cd project
$ docker run -it --rm --name=example \
  --mount type=bind,source=${PWD},target=/src \
  example/example_build:0.1 \
  bash
```

The parameter of particular importance in this context is --mount type=bind,source=$ {PWD},target=/src. This directive instructs Docker to bind mount the current directory, which houses the source code, to the /src directory within the container. This approach circumvents the need to copy source files into the container. Moreover, as will be demonstrated subsequently, it enables the storage of output binaries directly on the host's file system, thereby eliminating the need for redundant copying. For an understanding of the remaining flags and options, it is advisable to consult the official Docker documentation.

Within the container, we will proceed to compile the project:

```
root@3abec58c9774:/# cd src
root@3abec58c9774:/src# mkdir build && cd build
root@3abec58c9774:/src/build# cmake ..
-- The C compiler identification is GNU 7.5.0
-- The CXX compiler identification is GNU 7.5.0
-- Check for working C compiler: /usr/bin/cc
-- Check for working C compiler: /usr/bin/cc -- works
-- Detecting C compiler ABI info
-- Detecting C compiler ABI info - done
-- Detecting C compile features
-- Detecting C compile features - done
-- Check for working CXX compiler: /usr/bin/c++
-- Check for working CXX compiler: /usr/bin/c++ -- works
-- Detecting CXX compiler ABI info
-- Detecting CXX compiler ABI info - done
-- Detecting CXX compile features
-- Detecting CXX compile features - done
-- Boost  found.
-- Found Boost components:
   filesystem
-- Configuring done
-- Generating done
-- Build files have been written to: /src/build

root@3abec58c9774:/src/build# make
Scanning dependencies of target a.out
[ 50%] Building CXX object CMakeFiles/a.out.dir/main.cpp.o
[100%] Linking CXX executable a.out
[100%] Built target a.out
```

Et voilà, the project was built successfully!

The resulting binary runs successfully, both in the container and on the host, because Boost is linked *statically*:

```
$ build/a.out
The size of "/home/dima/dockerized_cpp_build_example/build/a.out" is
177320
```

Making the environment usable

At this juncture, it's reasonable to feel overwhelmed by the multitude of Docker commands and wonder how one is expected to memorize them all. It's important to emphasize that developers are not expected to retain every detail of these commands for project-building purposes. To streamline this process, I propose encapsulating the Docker commands within a tool that is widely familiar to most developers – make.

To facilitate this, I have established a GitHub repository (https://github.com/f-squirrel/dockerized_cpp) that contains a versatile Makefile. This Makefile is designed to be easily adaptable and can typically be employed for nearly any project that utilizes CMake without necessitating modifications. Users have the option to either directly download it from this repository or integrate it into their project as a Git submodule, ensuring access to the most recent updates. I advocate for the latter approach and will provide further details on this.

The Makefile is configured to support fundamental commands. Users can display the available command options by executing make help in the terminal:

```
$ make help
gen_cmake                     Generate cmake files, used internally
build                         Build source. In order to build a
specific target run: make TARGET=<target name>.
test                          Run all tests
clean                         Clean build directory
login                         Login to the container. Note: if the
container is already running, login into the existing one
build-docker-deps-image       Build the deps image.
```

To integrate the Makefile into our sample project, we'll begin by adding it as a Git submodule within the build_tools directory:

```
git submodule add  https://github.com/f-squirrel/dockerized_cpp.git
build_tools/
```

The next step is to create another Makefile in the root of the repository and include the Makefile that we have just checked out:

```
include build_tools/Makefile
```

Before the project compilation, it's prudent to adjust certain default settings to better suit the specific needs of your project. This can be efficiently achieved by declaring variables in the top-level Makefile prior to the inclusion of `build_tools/Makefile`. Such preemptive declarations allow for the customization of various parameters, ensuring that the build environment and process are optimally configured for your project's requirements:

```
PROJECT_NAME=example
DOCKER_DEPS_VERSION=0.1

include build_tools/Makefile
```
By defining the project name, we automatically set the build image name as `example/example_build`.

Make is now ready to build the image:

```
$ make build-docker-deps-image
docker build  -t example/example_build:latest \
 -f ./DockerfileBuildEnv .
Sending build context to Docker daemon  1.049MB
Step 1/6 : FROM ubuntu:18.04

< long output of docker build >

Build finished. Docker image name: "example/example_build:latest".
Before you push it to Docker Hub, please tag it(DOCKER_DEPS_VERSION +
1).
If you want the image to be the default, please update the following
variables:
/home/dima/dockerized_cpp_build_example/Makefile: DOCKER_DEPS_
VERSION
```

The Makefile, by default, assigns the latest tag to the Docker image. For better version control and to align with our project's current stage, it is advisable to tag the image with a specific version. In this context, we shall tag the image as 0.1.

Finally, let us build the project:

```
$ make
docker run -it --init --rm --memory-swap=-1 --ulimit core=-1
--name="example_build" --workdir=/example --mount type=bind,source=/
home/dima/dockerized_cpp_build_example,target=/example  example/
example_build:0.1 \
 bash -c \
 "mkdir -p /example/build && \
 cd build && \
 CC=clang CXX=clang++ \
```

```
  cmake   ..”
-- The C compiler identification is Clang 6.0.0
-- The CXX compiler identification is Clang 6.0.0
-- Check for working C compiler: /usr/bin/clang
-- Check for working C compiler: /usr/bin/clang -- works
-- Detecting C compiler ABI info
-- Detecting C compiler ABI info - done
-- Detecting C compile features
-- Detecting C compile features - done
-- Check for working CXX compiler: /usr/bin/clang++
-- Check for working CXX compiler: /usr/bin/clang++ -- works
-- Detecting CXX compiler ABI info
-- Detecting CXX compiler ABI info - done
-- Detecting CXX compile features
-- Detecting CXX compile features - done
-- Boost   found.
-- Found Boost components:
   filesystem
-- Configuring done
-- Generating done
-- Build files have been written to: /example/build


CMake finished.
docker run -it --init --rm --memory-swap=-1 --ulimit core=-1
--name="example_build" --workdir=/example --mount type=bind,source=/
home/dima/dockerized_cpp_build_example,target=/example   example/
example_build:latest \
 bash -c \
 "cd build && \
 make -j $(nproc) "
Scanning dependencies of target a.out
[ 50%] Building CXX object CMakeFiles/a.out.dir/main.cpp.o
[100%] Linking CXX executable a.out
[100%] Built target a.out


Build finished. The binaries are in /home/dima/dockerized_cpp_
build_example/build
```

Upon inspecting the build directory on the host, you'll observe that the output binary has been seamlessly placed there, facilitating easy access and management.

Both the Makefile and an example of a project that utilizes it with its default values can be found on GitHub. This provides a practical demonstration of how the Makefile can be integrated into a project, offering a turnkey solution for developers seeking to implement a Dockerized build environment in their C++ projects:

- Makefile repository: `https://github.com/f-squirrel/dockerized_cpp`
- Example project: `https://github.com/f-squirrel/dockerized_cpp`

Enhancements for user management within Dockerized builds

The initial iteration of the Docker-based build system executed operations under the root user's privileges. While this setup typically doesn't pose immediate problems—developers have the option to modify file permissions using `chmod`—executing Docker containers as the root user is generally discouraged from a security standpoint. More critically, this approach can lead to complications if any of the build targets modify the source code, such as code formatting or applying `clang-tidy` corrections through `make` commands. Such modifications could result in source files being owned by the root user, thereby restricting the ability to edit these files directly from the host.

To address this concern, modifications have been made to the Dockerized build's source code, enabling the container to execute as the host user by specifying the current user's ID and group ID. This adjustment is now the standard configuration to enhance security and usability. Should there be a need to revert to running the container as the root user, the following command can be utilized:

```
make DOCKER_USER_ROOT=ON
```

It is important to recognize that the Docker image does not replicate the host user's environment in its entirety—there is no corresponding home directory, nor are the user's name and group replicated within the container. This implies that if the build process relies on accessing the home directory, this modified approach may not be suitable.

Summary

In this chapter, we explored various strategies and tools for managing third-party dependencies in C++ projects, a critical aspect that significantly impacts the efficiency and reliability of the development process. We delved into traditional methods, such as utilizing operating system package managers and incorporating dependencies directly via Git submodules, each with its unique advantages and limitations.

We then transitioned to more specialized C++ package managers, highlighting Conan and vcpkg. Conan, with its robust ecosystem, extensive library support through Conan Center, and flexible configuration options, offers a comprehensive solution for managing complex dependencies, integrating seamlessly with multiple build systems, and supporting both static and dynamic linking. Its ability to handle multiple versions of libraries and the ease with which developers can extend the repository with custom packages make it an invaluable tool for modern C++ development.

vcpkg, developed by Microsoft, presents a slightly different approach, focusing on source-based distribution and ensuring libraries are built with the same compiler and settings as the consuming project. Its tight integration with CMake and Visual Studio, coupled with the backing of Microsoft, ensures a smooth experience, particularly for projects within the Microsoft ecosystem. The emphasis on compiling from source addresses potential binary incompatibility issues, making vcpkg a reliable choice for managing dependencies.

Lastly, we discussed the adoption of Dockerized builds as an advanced strategy for creating consistent, reproducible build environments, particularly beneficial in Linux systems. This approach, while more complex, offers significant advantages in terms of isolation, scalability, and consistency across development, testing, and deployment stages.

Throughout the chapter, we aimed to equip you with the knowledge and tools necessary to navigate the landscape of dependency management in C++ projects. By understanding the strengths and limitations of each method and tool, developers can make informed decisions tailored to their project's specific needs, leading to more efficient and reliable software development processes.

14
Version Control

In software development, maintaining a clean commit history is essential for producing enduring and coherent code. This chapter emphasizes that a well-organized commit history is fundamental to robust software engineering. By focusing on version control, particularly through clear commit summaries and messages, we will explore the techniques and intentional practices needed to achieve clarity and precision.

Committing code is like adding individual threads to the overall narrative of a project's development. Each commit, with its summary and message, contributes to the understanding of the project's history and future direction. Maintaining a clean commit history goes beyond organizational neatness; it embodies effective communication among developers, facilitates seamless collaboration, and enables quick navigation through the project's development history.

In the following sections, we will examine what makes a "good" commit, focusing on attributes that bring clarity, purpose, and utility to commit messages. This exploration goes beyond the basics, delving into strategic documentation of code changes and insights gained through tools such as Git. With illustrative examples, we will see how well-crafted commit histories can transform understanding, aid in debugging, and streamline the review process by clearly conveying the rationale behind code alterations.

Advancing further, we will decode the Conventional Commits specification, a structured framework designed to standardize commit messages, thereby infusing them with predictability and machine-parseable clarity. This section illuminates the symbiotic relationship between commit message structure and automated tooling, showcasing how adherence to such conventions can dramatically enhance project maintainability.

As we progress, the narrative unfolds to reveal the practicalities of enforcing these best practices through the lens of commit linting. Here, we delve into the integration of automated tools within **Continuous Integration** (CI) workflows, demonstrating how such mechanisms serve as vigilant guardians of commit quality, ensuring consistency and compliance with established norms.

This chapter goes beyond explaining the mechanics of version control; it invites you to view crafting clean commit histories as a vital part of software craftsmanship. By following the principles and practices discussed here, developers and teams can improve their code repositories' quality and create an environment that promotes innovation, collaboration, and efficiency. As we explore this chapter, remember that a clear commit history reflects our dedication to excellence in software development.

What is a good commit?

At the heart of effective version control practices lies the concept of a "good commit," a fundamental unit of change that embodies the principles of clarity, atomicity, and purposefulness within the code base. Understanding what constitutes a good commit is essential for developers who strive to maintain a clean, navigable, and informative project history. This section delves into the key attributes that define the quality of a commit, offering insights into how developers can enhance their version control practices.

The principle of singular focus

A good commit adheres to the principle of atomicity, meaning it encapsulates a single logical change within the code base. This singular focus ensures that each commit is independently meaningful and that the project can be safely and easily reverted or modified by reverting or adjusting individual commits. Atomic commits simplify code review processes, making it easier for team members to understand and evaluate each change without the noise of unrelated modifications. For example, instead of combining a new feature implementation with a separate bug fix in one commit, they should be split into two distinct commits, each with its clear purpose and scope.

The art of communication

The essence of a good commit also lies in its clarity, particularly evident in the commit message. A clear commit message succinctly describes the what and the why of the change, serving as concise documentation for future reference. This clarity extends beyond the immediate team, aiding anyone who interacts with the code base, including new team members, external collaborators, and even the future self. This becomes especially important when revisiting the code base after a prolonged period, as the commit messages serve as a historical record of the project's evolution. This approach is crucial for open source projects because it allows contributors to understand the context and rationale behind the changes, thereby fostering a collaborative and inclusive environment.

A well-structured commit message typically includes a concise title line summarizing the change, followed by a blank line and a more detailed explanation if necessary. The explanation can delve into the rationale behind the change, any implications it might have, and any additional context that helps understand the commit's purpose. It is recommended to keep the subject line up to 50 characters. This ensures that the message fits within the standard width of most terminals, is not terminated by GitHub or other platforms, and is easily scannable. GitHub truncates subjects shorter than 72 characters, so 72 will be a hard limit and 50 a soft one. For example, the commit message `feat: added a lots of needed include directives to make things compile properly` will be truncated by GitHub as follows:

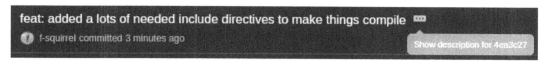

Figure 14.1 – Truncated commit message

GitHub truncates the last word, `properly`, and in order to read it, developers will have to click on the commit message. This is not a big deal but it is a small inconvenience that can be easily avoided by keeping the subject line short.

More importantly, it forces the author to be concise and to the point.

Other useful practices include using the imperative mood in the subject line, which is a common convention in commit messages. This means that the subject line should be phrased as a command or instruction, such as `fix the bug` or `add the feature`. This style of writing is more direct and aligns with the idea that a commit represents a change that is being applied to the code base.

Prefer not to end the subject line with a period, as it is not a complete sentence and does not help to keep the message short. The body of the commit message can provide additional context, such as the motivation for the change, the problem it solves, and any relevant details about the implementation.

Prefer to wrap the body at 72 characters, because Git does not wrap text for you. It is a common convention and it makes the message more readable in various contexts, such as terminal windows, text editors, and version control tools. It can be easily achieved by configuring your code editor.

Therefore, the moment spent in reflection before finalizing a commit is not just about ensuring the clarity of the message; it is about reaffirming the value and intent behind the changes themselves. It is an opportunity to ensure that each contribution to the code base is deliberate, meaningful, and aligned with the project's objectives. In this light, taking the time to craft a precise and informative commit message is not only a good practice but a testament to the developer's commitment to quality and collaboration.

The art of refinement

Before merging feature branches into the `main` branch, it's prudent for developers to consider the cleanliness and clarity of their commit history. Squashing intermediate commits is a thoughtful practice that streamlines the commit log, making it more readable and meaningful for anyone who explores the project history.

When you're on the verge of integrating your work, take a moment to reflect on the commit messages that have accumulated during development. Ask yourself whether each commit message adds value to the understanding of the project's evolution or whether it merely clutters the history with redundant or overly granular details. In many cases, the iterative steps you took to arrive at the final solution— such as minor bug fixes, adjustments in response to code reviews, or corrections to unit tests—may not hold significant value for other contributors or for the future you.

Consider a commit history filled with messages such as `fix bug`, `fix unit test`, or multiple `fix cr` entries. Such messages, while indicative of the development process, do not necessarily provide meaningful insights into the changes or their impact on the project. Squashing these intermediate commits into a single, well-crafted commit not only tidies up the commit log but also ensures that each entry in the history conveys a significant step in the project's development.

By squashing commits, you consolidate these iterative changes into a cohesive narrative that highlights the introduction of a new feature, the resolution of a significant bug, or the implementation of a crucial refactor. This curated history aids both current contributors and future maintainers in navigating and understanding the project's progression, enhancing collaboration and efficiency.

In summary, before merging, consider the broader perspective of the project's commit history. Squashing intermediate commits is a practice in mindfulness, ensuring that the commit log remains a valuable and navigable resource for all contributors, encapsulating the essence of each change in a clear and concise manner.

Conventional Commits specification

Consistency in commit messages and structures across a project enhances readability and predictability, making it easier for team members to navigate the project history. Adhering to a predefined format or set of conventions, such as the Conventional Commits specification, ensures that commit messages are uniformly structured and informative. These might include starting commit messages with a verb in the imperative mood, specifying the type of change (e.g., `fix`, `feat`, or `refactor`), and optionally including a scope to clarify what part of the project is affected.

Linking code to context

A good commit enhances traceability by linking the code change to its broader context, such as a ticket in the project's issue-tracking system or relevant documentation. Including references in commit messages creates a tangible connection between the technical implementation and the requirements or issues it addresses, facilitating better understanding and tracking of project progress.

Incorporating issue tracker IDs, ticket numbers, or other relevant identifiers in commit messages can significantly improve the traceability of changes and the ease with which they can be related to the project's objectives or reported issues. It often looks similar to `fix(FP-1234): corrected the user authentication flow`, where `FP-1234` is the ticket number in the issue-tracking system.

In essence, a good commit acts as a coherent, self-contained story within the broader narrative of a project's development history. By adhering to these principles, developers not only contribute to the maintainability and readability of the code base but also foster a culture of meticulousness and accountability in version control practices. Through the disciplined creation of good commits, the project's history becomes a valuable asset for collaboration, review, and understanding the evolution of the software.

One of the best ways to create good commit messages is by following the Conventional Commits specification. The Conventional Commits specification stands as a structured framework for commit message formatting, designed with the dual aim of simplifying the process of creating readable commit logs and enabling automated tooling to facilitate version management and release note generation. This specification delineates a standardized format for commit messages, intending to clearly communicate the nature and intent of changes within a version control system, such as Git.

Overview and intent

At its core, the Conventional Commits specification prescribes a format that includes a type, an optional scope, and a succinct description. The format typically follows this structure:

```
<type>[optional scope]: <description>

[optional body]

[optional footer(s)]
```

The *type* categorizes the commit according to the nature of the change it introduces, such as `feat` for a new feature or `fix` for a bug fix. The *scope*, though optional, provides additional contextual information, often indicating the part of the code base affected by the change. The *description* offers a concise summary of the change, crafted in an imperative mood.

Options and usage

Commit linting, particularly when adhering to the Conventional Commits specification, ensures that commits are structured in a clear, predictable, and useful manner. Here are some examples of commits that comply with commit linting rules, showcasing various types of changes that might occur in a software project:

- Adding a new feature:

  ```
  feat(authentication): add biometric authentication support
  ```

  This commit message indicates that a new feature (`feat`) has been added, (specifically, biometric authentication support) and the scope of this feature is within the authentication module of the application.

- Fixing a bug:

  ```
  fix(database): resolve race condition in user data retrieval
  ```

  Here, a bug fix (`fix`) is being committed, addressing a race condition issue within the `database` module, particularly in the process of user data retrieval.

- Improving documentation:

```
docs(readme): update installation instructions
```

This example shows a documentation update (docs), with the changes made to the project's README file to update the installation instructions.

- Code refactoring:

```
refactor(ui): simplify button component logic
```

In this commit, existing code has been refactored (refactor) without adding any new features or fixing any bugs. The scope of the refactoring is the UI, specifically simplifying the logic used in a button component.

- Style adjustments:

```
style(css): remove unused CSS classes
```

This commit message signifies a styling change (style), where unused CSS classes are being removed. It's worth noting that this type of commit does not affect the functionality of the code.

- Adding tests:

```
test(api): add tests for new user endpoint
```

Here, new tests (test) have been added for a new user endpoint in the API, indicating an enhancement in the project's test coverage.

- Chore tasks:

```
chore(build): update build script for deployment
```

This commit represents a chore (chore), typically a maintenance or setup task that doesn't directly modify the source code or add functionality, such as updating a build script for deployment.

- Breaking change:

```
feat(database): change database schema for users table

BREAKING CHANGE: The database schema modification requires
resetting the database. This change will affect all services
interacting with the users table.
```

Another method to indicate a breaking change is by adding an exclamation mark (!) after the type and scope but before the colon in the commit message. This method is succinct and visually noticeable:

```
feat!(api): overhaul authentication system
```

This commit introduces a new feature (feat) related to the database schema of the user's table but also includes a breaking change.

These examples illustrate how commit linting, guided by the Conventional Commits specification, facilitates clear, structured, and informative commit messages that enhance project maintainability and collaboration.

Origins and adoption

The Conventional Commits specification was inspired by the need to streamline the creation of readable and automated changelogs. It builds upon earlier practices from the AngularJS team and has since been adopted by various open source and enterprise projects seeking to standardize commit messaging for improved project maintainability and collaboration.

Advantages of Conventional Commits

Adhering to the Conventional Commits specification offers numerous benefits:

- **Automated Semver handling**: By categorizing commits, tools can automatically determine version bumps based on the semantic meaning of changes, adhering to **Semantic Versioning (Semver)** principles

- **Streamlined release notes**: Automated tools can generate comprehensive and clear release notes and changelogs by parsing structured commit messages, significantly reducing manual effort and enhancing release documentation

- **Enhanced readability**: The standardized format improves the readability of commit history, making it easier for developers to navigate and understand project evolution

- **Facilitated code reviews**: The clear categorization and description of changes aid in the code review process, enabling reviewers to quickly grasp the scope and intent of changes

Commitlint – enforcing commit message standards

Commitlint is a powerful, configurable tool designed to enforce commit message conventions, ensuring consistency and clarity across a project's commit history. It plays a crucial role in maintaining a clean, readable, and meaningful commit log, particularly when used in conjunction with conventions such as the Conventional Commits specification. This section provides a comprehensive guide on how to install, configure, and use commitlint to check commit messages locally, fostering a disciplined approach to version control and collaboration.

Installation

Commitlint is typically installed via npm, the package manager for Node.js. To get started, you'll need to have Node.js and npm installed on your development machine. Once set up, you can install commitlint and its conventional `config` package by running the following commands in your project's root directory:

```
npm install --save-dev @commitlint/{cli,config-conventional,prompt-cli}
```

This command installs commitlint and the conventional commits configuration as development dependencies in your project, making them available for use in the local development environment.

Configuration

After installation, commitlint requires a configuration file to define the rules it will enforce. The most straightforward way to configure commitlint is by using the conventional commits configuration, which aligns with the Conventional Commits specification. Create a file named `commitlint.config.js` in your project's root directory, and add the following content:

```
module.exports = {extends: ['@commitlint/config-conventional']};
```

This configuration instructs commitlint to use the standard rules provided by the Conventional Commits configuration, which include checks for the structure, types, and scopes of commit messages.

Local usage

To check commit messages locally, you can use commitlint in conjunction with Husky, a tool for managing Git hooks. Husky can be configured to trigger commitlint to evaluate commit messages before they are committed, providing immediate feedback to the developer.

First, install Husky as a development dependency:

```
npm install --save-dev husky
```

Let us check local commits with commitlint:

```
npx commitlint --from HEAD~1 --to HEAD --verbose
✗   input: feat: add output
✓   found 0 problems, 0 warnings
```

In this example, `--from HEAD~1` and `--to HEAD` specify the range of commits to check, and `--verbose` provides detailed output. If the commit message does not adhere to the specified conventions, commitlint will output an error message, indicating the violations that need to be addressed.

Let us add a bad commit message and check it with commitlint:

```
git commit -m"Changed output type"
npx commitlint --from HEAD~1 --to HEAD --verbose
ⅈ    input: Changed output type
✖    subject may not be empty [subject-empty]
✖    type may not be empty [type-empty]

✖    found 2 problems, 0 warnings
ⅈ    Get help: https://github.com/conventional-changelog/
commitlint/#what-is-commitlint
```

Commitlint can be integrated as a Git hook by adding the following configuration to your `package.`
`json` file or by creating a `.huskyrc` file with the same content:

```
"husky": {
  "hooks": {
    "commit-msg": "commitlint -E HUSKY_GIT_PARAMS"
  }
}
```

This configuration sets up a pre-commit hook that invokes commitlint with the commit message that is about to be committed. If the commit message does not meet the specified standards, commitlint will reject the commit, and the developer will need to revise the message accordingly.

Customizing rules

Commitlint offers a wide array of configuration and customization options, allowing teams to tailor commit message validation rules to fit their specific project requirements and workflows. This flexibility ensures that commitlint can be adapted to support various commit conventions beyond the standard Conventional Commits format, providing a robust framework for enforcing consistent and meaningful commit messages across diverse development environments.

Basic configuration

The basic configuration of commitlint involves setting up a `commitlint.config.js` file in your project's root directory. This file serves as the central point for defining the rules and conventions that commitlint will enforce. At its simplest, the configuration might extend a predefined set of rules, such as those provided by `@commitlint/config-conventional`, as shown here:

```
module.exports = {
  extends: ['@commitlint/config-conventional'],
};
```

This configuration instructs commitlint to use the conventional commit message rules, enforcing a standard structure and set of types for commit messages.

Custom rule configuration

Commitlint's real power lies in its ability to customize rules to match specific project needs. Each rule in commitlint is identified by a string key and can be configured with an array specifying the rule's level, applicability, and, in some cases, additional options or values. The rule configuration array generally follows the format [level, applicability, value]:

- **Level**: Determines the severity of violations (0 = disabled, 1 = warning, and 2 = error)

- **Applicability**: Specifies whether the rule is applied ('always' or 'never')

- **Value**: Additional parameters or options for the rule, varying by rule

For example, to enforce that commit messages must start with a type followed by a colon and a space, you could configure the type-enum rule as follows:

```
module.exports = {
  rules: {
    'type-enum': [2, 'always', ['feat', 'fix', 'docs', 'style',
'refactor', 'test', 'chore']],
  },
};
```

This configuration sets the rule level to error (2), specifies that the rule should always be applied, and defines a list of acceptable types for commit messages.

Scope and subject configuration

Commitlint also allows for detailed configuration of commit message scopes and subjects. For instance, you can enforce specific scopes or require that commit message subjects do not end with a period:

```
module.exports = {
  rules: {
    'scope-enum': [2, 'always', ['ui', 'backend', 'api', 'docs']],
    'subject-full-stop': [2, 'never', '.'],
  },
};
```

This setup mandates that commits must use one of the predefined scopes and that the subject line must not end with a period.

Customizing and sharing configurations

For projects or organizations with unique commit message conventions, custom configurations can be defined and, if needed, shared across multiple projects. You can create a dedicated npm package for your commitlint configuration, allowing teams to easily extend this shared configuration:

```
// commitlint-config-myorg.js
module.exports = {
  rules: {
    // Custom rules here
  },
};

// In a project's commitlint.config.js
module.exports = {
  extends: ['commitlint-config-myorg'],
};
```

This approach promotes consistency across projects and simplifies the management of commit message rules within an organization.

Integration with CI

Ensuring the enforcement of commitlint via CI is a crucial practice in maintaining high-quality commit messages across a project. While local Git hooks, such as those managed by Husky, offer a first line of defense by checking commit messages on the developer's machine, they are not infallible. Developers might intentionally or accidentally disable Git hooks, and **Integrated Development Environments (IDEs)** or text editors might not be properly configured to enforce these hooks or might encounter issues that lead to their malfunction.

Given these potential gaps in local enforcement, CI serves as the authoritative source of truth, providing a centralized, consistent platform for verifying commit messages against the project's standards. By integrating commitlint into the CI pipeline, projects ensure that every commit, regardless of its origin or the method used to submit it, adheres to the defined commit message conventions before it is merged into the main code base. This CI-based enforcement fosters a culture of discipline and accountability, ensuring that all contributions, regardless of their source, meet the project's quality standards.

Integrating commitlint into CI with GitHub Actions

GitHub Actions offers a straightforward and powerful platform for integrating commitlint into your CI workflow. The following example demonstrates how to set up a GitHub action to enforce commit message standards using commitlint on every push or pull request targeting the main branch.

First, create a new file in your repository under .github/workflows/commitlint.yml with the following content:

```
name: Commitlint
on:
  push:
    branches: [ main ]
  pull_request:
    branches: [ main ]

jobs:
  commitlint:
    runs-on: ubuntu-latest
    steps:
    - name: Check out code
      uses: actions/checkout@v3
      with:
        # Fetch at least the immediate parents so that if this is
        # a pull request then we can checkout the head.
        fetch-depth: 0

    - name: Check Commit Message
      uses: wagoid/commitlint-github-action@v5
      with:
        failOnWarnings: true
```

This workflow defines a job named commitlint that triggers on pushes and pull requests to the main branch. The only configuration that I would like to highlight is failOnWarnings: true, which configures the action to fail the job if any commitlint warnings are encountered. This ensures strict adherence to the commit message standards by treating warnings with the same severity as errors.

Let us create a bad commit message and open a PR to see how the action works:

```
git commit -m"Changed output type"
git checkout -b exit_message
git push origin exit_message
```

After we open a PR, we will see that the action has failed:

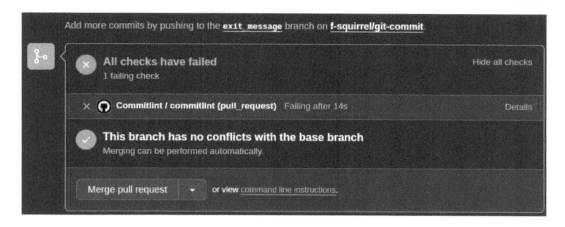

Figure 14.2 – Commitlint action failed

The logs will show the reason for the failure in the same format as the local check:

```
9   Error: You have commit messages with errors
10
11  x    input: Changed output type
12  *    subject may not be empty [subject-empty]
13  *    type may not be empty [type-empty]
14
15  *    found 2 problems, 0 warnings
16  ⓘ    Get help: https://github.com/conventional-changelog/commitlint/#what-is-commitlint
17
```

Figure 14.3 – Commitlint action log failed

By incorporating this workflow into your project, you ensure that every commit is scrutinized for adherence to your commit message standards before it becomes part of the main branch. This CI-based check acts as a final gatekeeper, reinforcing the importance of well-structured commit messages and maintaining the integrity of the project's commit history.

Commitlint's configurability and customization options provide a powerful platform for enforcing commit message standards tailored to a project's or organization's specific needs. By leveraging these capabilities, teams can ensure that their commit logs remain clear, consistent, and meaningful, thereby enhancing project maintainability and collaboration. Whether adhering to widely accepted conventions such as the Conventional Commits specification or defining a set of custom rules, commitlint offers the flexibility and control needed to maintain a high-quality commit history.

Generating changelogs

Automatic changelog generation is a method where software tools automatically create a log of changes made to a project, categorizing and listing updates, fixes, and features. This process is favored for its efficiency and consistency, ensuring that all significant modifications are documented systematically. We'll explore this concept through GitCliff, a tool that parses structured commit messages to generate detailed changelogs, aiding in transparent project management and communication. GitCliff's utility in this process exemplifies its role in automating and streamlining project documentation tasks.

Installation

GitCliff is written in Rust and can be installed using Cargo, the Rust package manager. To install GitCliff, ensure that you have Rust and Cargo installed on your system, and then run the following command:

```
curl https://sh.rustup.rs -sSf | sh
```

After installing Rust, you can install GitCliff using Cargo:

```
cargo install git-cliff
```

The last configuration step is to initialize GitCliff in your project:

```
git cliff --init
```

This generates a default configuration file, `.cliff.toml`, in the root of your project.

GitCliff usage

After installing and initializing GitCliff, you can generate a changelog by running the following command in your project's root directory:

```
git cliff -o CHANGELOG.md
```

The tool generates a Markdown file with the changelog, as follows:

Changelog

All notable changes to this project will be documented in this file.

[unreleased]

🚀 Features

- Initial commit
- Add output
- Change output text
- [**breaking**] Change CI parameters

❄ Miscellaneous Tasks

- *(ci)* Add commitlint config
- *(git)* Ignore node_modules
- *(ci)* Add commitlint
- *(git)* Add git-cliff

Figure 14.4 – Generated changelog

The log contains a list of changes, categorized by type, and it highlights breaking changes.

Let us add a release tag and generate a changelog for the release:

```
git tag v1.0.0 HEAD
git cliff
```

The changelog will now contain the release tag and the changes since the last release:

Changelog

All notable changes to this project will be documented in this file.

[0.1.0] - 2024-03-16

🚀 Features

- Initial commit
- Add output
- Change output text
- [**breaking**] Change CI parameters

🌼 Miscellaneous Tasks

- *(ci)* Add commitlint config
- *(git)* Ignore node_modules
- *(ci)* Add commitlint
- *(git)* Add git-cliff

Figure 14.5 – Generated changelog with release tag

We can introduce a breaking change and bump the version:

```
git commit -m"feat!: make breaking change"
git cliff --bump
```

As you can see, GitCliff has detected the breaking change and bumped the version to 2.0.0:

Changelog

All notable changes to this project will be documented in this file.

[2.0.0] - 2024-03-23

🚀 Features

- [**breaking**] Make breaking change

[1.0.0] - 2024-03-16

🚀 Features

- Initial commit
- Add output
- Change output text
- [**breaking**] Change CI parameters

⚙️ Miscellaneous Tasks

- *(ci)* Add commitlint config
- *(git)* Ignore node_modules
- *(ci)* Add commitlint
- *(git)* Add git-cliff

Figure 14.6 – Generated changelog with breaking change

In the preceding sections, we have comprehensively explored the significant functionalities of git-cliff, revealing its substantial utility in automating change log generation. This tool distinguishes itself not only through its capacity to streamline the documentation process but also through its seamless integration with CI platforms, including but not limited to GitHub. Such integration ensures that changelogs are consistently synchronized with the latest project developments, thereby maintaining the accuracy and relevance of project documentation.

An equally noteworthy feature of git-cliff is the extensive customization it offers for changelog generation. Users are afforded the flexibility to tailor the format, content, and presentation of changelogs to meet specific project requirements or personal preferences. This high degree of customizability ensures that the output not only aligns with but also enhances the project's documentation standards.

Given the depth of functionality and the potential benefits that git-cliff offers, those interested in leveraging this tool to its fullest are encouraged to consult the official documentation. This resource is a treasure trove of detailed information, covering the breadth of features, configurations, and best practices associated with git-cliff. Engaging with the official documentation will not only solidify your understanding of the tool but also equip you with the knowledge to implement it effectively in your projects.

To summarize, having delved into the major capabilities and advantages of git-cliff, the path forward for those looking to integrate this tool into their development workflow is through a thorough exploration of the official documentation. This exploration promises to extend your proficiency in utilizing git-cliff, ensuring that you can fully harness its capabilities to enhance your project's change log generation and documentation processes.

Utilizing git-bisect in bug hunting

During the process of software development, the task of identifying and rectifying bugs is paramount to ensure the stability and reliability of the application. Among the arsenal available to developers for this purpose, git-bisect stands out as a powerful tool, specifically designed for the task of isolating the commit that introduced a bug into the code base.

Embedded within the Git version control system, git-bisect is a utility based on the binary search algorithm. It aids developers in sifting through a potentially vast commit history to pinpoint the exact change that caused a regression or introduced an error. By adopting a divide-and-conquer strategy, git-bisect significantly streamlines the debugging process, making it an efficient approach to troubleshooting.

The journey with git-bisect begins by establishing two critical points in the project's timeline: a commit where the bug is known to be absent (referred to as good) and a commit where the bug is confirmed to be present (bad). With these markers set, git-bisect proceeds to check out a commit that lies midway between the good and bad commits. This step requires the developer to test the current state of the application to determine whether the bug is present or not.

The iterative process involves git-bisect selecting a new commit based on the developer's feedback, and continually narrowing down the search area by halving it with each step. The cycle of testing and feedback continues until git-bisect successfully isolates the commit that introduced the bug, effectively zeroing in on the root cause with minimal manual review.

The efficiency of git-bisect lies in its ability to reduce the number of commits that need to be scrutinized manually, thus conserving valuable development time. Its methodical approach ensures precision in identifying the problematic commit, which is crucial for understanding the context of the bug and formulating an effective fix. Being an integral part of the Git ecosystem, git-bisect seamlessly fits into the developer's existing workflow, offering a familiar and straightforward interface for debugging.

To optimize the effectiveness of `git-bisect`, it is imperative to use a reliable and accurate test case for evaluating each commit. This ensures that the feedback provided to `git-bisect` correctly reflects the presence or absence of the bug, thereby preventing misidentification. Maintaining a clean and logical commit history, where each commit encapsulates a single change, enhances the tool's efficiency. Furthermore, automating the testing process within the `git-bisect` session, when feasible, can expedite the bug-hunting endeavor.

Consider a scenario where a regression is detected in a feature that was previously functioning correctly. It happens often when certain tests are running only at night. The task is to identify the commit responsible for this regression using `git-bisect`:

1. Start the `bisect` session with `git bisect start`.

2. Mark the commit in which the bug is present as `git bisect bad <commit hash>` (usually, it is HEAD).

3. Identify a commit in the past where the feature worked correctly and mark it as good using `git bisect good <commit-hash>`.

4. `git-bisect` will then check out a commit halfway between the good and bad commits. Test this commit to see whether the bug exists.

5. Based on the test outcome, mark the commit as good or bad. `git-bisect` uses this feedback to narrow down the search space and selects a new commit for testing.

6. Repeat the testing and marking process until `git-bisect` identifies the commit that introduced the bug.

Once the problematic commit is identified, developers can examine the changes introduced in that commit to understand the cause of the bug and proceed with developing a fix.

To demonstrate how it works, I cloned the master branch of the `rapidjson` library a bug and put it in the middle of the local repo. The Git log looks as follows, where `Bad commit (6 hours ago) <f-squirrel>` is the bad one:

```
a85e2979 - (HEAD -> master) Add RAPIDJSON_BUILD_CXX20 option (6 hours
ago) <Brian Rogers>
2cd6149d - fix Visual Studio 2022 (using /std:c++20) warning warning
C5232: in C++20 this comparison ...
478cd636 - Bad commit (6 hours ago) <f-squirrel>
25edb27a - tests: Only run valgrind tests if valgrind was found (23
hours ago) <Richard W.M. Jones>
606791f6 - Fix static_cast in regex.h (23 hours ago) <Dylan Burr>
5f071d72 - Fix comparision of two doubles (23 hours ago) <Esther Wang>
060a09a1 - Fix schema regex preprocessor include logic (6 weeks ago)
<Bryant Ferguson>
6089180e - Use correct format for printf (4 months ago) <Esther Wang>
...
```

Let us start bisecting by marking good and bad commits:

```
$ git bisect start
$ git bisect bad HEAD
$ git bisect good 6089180e

Bisecting: 3 revisions left to test after this (roughly 2 steps)
[606791f6662c136ba34f842313b807114580852d] Fix static_cast in regex.h
```

I have prepared a script that checks whether the bug is present in the current commit. The script is called test.sh and looks as follows:

```
cmake --build ./build -j $(nproc) || exit 1
./build/bin/unittest || exit 1
```

Every time I run the script, I mark the commit as good or bad. After a few iterations, I have found the commit that introduced the bug:

```
[  PASSED  ] 468 tests.
$ git bisect good

Bisecting: 1 revision left to test after this (roughly 1 step)
[478cd636a813abe76e32154544b0ec793fdc5566] Bad commit
```

If we run the test script again, we will see that the bug is present in the commit:

```
[  FAILED  ] 2 tests, listed below:
[  FAILED  ] BigInteger.Constructor
[  FAILED  ] BigInteger.LeftShift

 2 FAILED TESTS
```

Once we finish our bug hunt, we can reset the bisect session with git bisect reset.

Jumping between commits for a user is a useful functionality but not the only one. git-bisect can be automated with a script that will run the tests and mark the commits as good or bad based on the test results. Note that the script should return 0 if the commit is good and 1 if the commit is bad. The script will run until the bug is found and the bisect session will be reset. For our repository, it will look as follows:

```
$ git bisect start
$ git bisect bad HEAD
$ git bisect good 6089180e

Bisecting: 3 revisions left to test after this (roughly 2 steps)
[606791f6662c136ba34f842313b807114580852d] Fix static_cast in regex.h
```

```
$ git bisect run ./test.sh

running './test.sh'

... build and test output ...
[==========] 468 tests from 34 test suites ran. (321 ms total)
[  PASSED  ] 468 tests.
478cd636a813abe76e32154544b0ec793fdc5566 is the first bad commit
commit 478cd636a813abe76e32154544b0ec793fdc5566
Author: f-squirrel <dmitry.b.danilov@gmail.com>
Date:   Mon Mar 25 15:18:18 2024 +0200

    Bad commit

 include/rapidjson/internal/biginteger.h | 2 +-
 1 file changed, 1 insertion(+), 1 deletion(-)
bisect found first bad commit
```

`git-bisect` is an indispensable debugging tool within the Git suite, offering a systematic and efficient approach to identifying bug-inducing commits. Its integration into the development workflow, combined with the practice of maintaining a clear commit history and employing automated tests, makes it a highly effective solution for maintaining code quality and stability.

Summary

In this chapter dedicated to version control, we embarked on an insightful journey through the core principles and practices that underpin effective software version management. Central to our exploration was the adoption of conventional commits, a structured approach to commit messaging that enhances readability and facilitates automated processing of commit logs. This practice, grounded in a standardized format for commit messages, enables teams to convey the nature and intent of changes with clarity and precision.

We also delved into SemVer, a methodology designed to manage version numbers in a meaningful way. SemVer's systematic approach to versioning, based on the significance of changes in the code base, provides clear guidelines on when and how version numbers should be incremented. This method offers a transparent framework for version control, ensuring compatibility and facilitating effective dependency management within and across projects.

The chapter further introduced change log creation tools, with a particular focus on git-cliff, a versatile tool that automates the generation of detailed and customizable change logs from Git history. These tools streamline the documentation process, ensuring that project stakeholders are well informed about the changes, features, and fixes introduced with each new version.

A significant portion of the chapter was dedicated to debugging techniques, highlighting the utility of `git-bisect` in the process of isolating bugs. `git-bisect`, through its binary search algorithm, enables developers to efficiently pinpoint the commit that introduced a bug, thereby significantly reducing the time and effort required for troubleshooting.

In summary, this chapter provided a comprehensive overview of version control practices, emphasizing the importance of structured commit messages, strategic versioning, automated changelog generation, and efficient debugging techniques. By adopting these practices, development teams can enhance collaboration, maintain code base integrity, and ensure the delivery of high-quality software.

In the next chapter, we will turn our attention to a critical aspect of the development process: code reviews. We will examine the importance of code reviews in ensuring code quality, fostering team collaboration, and enhancing overall productivity. By understanding best practices and effective strategies for conducting thorough and constructive code reviews, you will be well equipped to elevate the standards of your code base and contribute more effectively to your team's success. Stay tuned as we embark on this insightful journey into the art and science of code reviews.

15
Code Review

In the preceding chapters of this book, we have systematically explored a range of automated solutions designed to enhance the quality of our C++ code. These include the adoption of clear naming conventions, the utilization of modern C++ features, the rigorous implementation of unit and end-to-end testing, the maintenance of clean commits and messages, and the effective use of debugging techniques such as git bisect, among others. Each of these practices serves as a vital component in our toolkit for writing robust, maintainable software.

However, despite the substantial benefits conferred by these automated tools and methodologies, they are not infallible. They rely on correct and consistent implementation, and without diligent oversight, it's all too easy for standards to slip and errors to creep into our codebase. This is where the critical role of code review comes into play. A human eye, capable of interpreting context and nuance, is essential in ensuring that all these automated practices are correctly and effectively applied.

In this chapter, we will delve deep into the practice of code review, an indispensable part of the development process that helps safeguard against the oversights that no machine can yet fully detect. We'll discuss how code reviews not only prevent potential bugs and enhance code quality but also foster a collaborative culture of learning and accountability among developers. Through a comprehensive exploration of effective strategies and practical guidelines, we aim to equip you with the knowledge to implement robust code reviews that significantly contribute to the success and reliability of your C++ projects.

What is a code review and why is it needed?

The practice of code review, as we understand it today, owes its origins to Michael Fagan, who developed the formal process of software inspection in the mid-1970s. At that time, software engineering was often a solitary pursuit, with individual developers acting as lone cowboys who were responsible for writing, testing, and reviewing their own code. This method led to inconsistent standards across projects and a higher incidence of overlooked errors, as individual biases and blind spots went unchecked.

Recognizing the limitations of this solitary approach, Fagan introduced a structured method to inspect software systematically. His process was aimed not only at finding errors but also at examining the overall design and implementation of software. This shift marked a significant evolution in software development, emphasizing collaboration, meticulous examination, and shared responsibility. By involving multiple reviewers, Fagan's method brought diverse perspectives to the evaluation process, enhancing the scrutiny and overall quality of the software.

Benefits of code reviews

Code reviews significantly enhance the overall quality of code by ensuring that the software's design is consistent with the project's architectural standards and the best practices of C++. These reviews are crucial for enforcing coding standards and conventions, thereby fostering a codebase that is more uniform and easier to comprehend for both new and existing team members. Moreover, they aid in maintaining high levels of understandability by facilitating discussions about complex sections of the code and clarifying the rationale behind specific approaches. For instance, consider a scenario where a developer employs unconventional loop structures that, although functional, are challenging to understand and maintain. During a code review, these issues can be brought to light, and a suggestion to refactor the code using standard STL algorithms might be made. This not only simplifies the code but also ensures alignment with modern C++ practices, enhancing readability and maintainability.

Peer reviews are one of the most effective ways to catch bugs early before they make it into production. It is always better to have another pair of eyes on the code to look for errors, whether they are logical mistakes or incorrect usage of the language. Reviewers can identify logic errors, off-by-one errors, memory leaks, and other common C++ pitfalls that might not be immediately apparent to the original author. Moreover, conducting a thorough review of test cases in unit tests during the code review process is equally vital. This practice ensures that the tests cover sufficient scenarios and catch potential bugs at an early stage, thereby enhancing the reliability and robustness of the software. For example, a developer might forget to free the memory allocated during a function, potentially causing a memory leak. A peer reviewing this code might spot this oversight and suggest the use of smart pointers to automatically manage the memory lifecycle, effectively preventing such issues before the software progresses further in the development cycle.

Reviews are invaluable as an educational tool, benefiting both the authors and the reviewers by disseminating domain knowledge and enhancing familiarity with the codebase throughout the team. This aspect of knowledge transfer is crucial in ensuring that all team members are aligned and capable of contributing effectively. For instance, a junior developer might initially use raw pointers to manage resources, which is a common practice yet prone to errors such as memory leaks and pointer-related bugs. During a code review, a more experienced developer can guide the junior by introducing them to smart pointers. By explaining the advantages of smart pointers, such as automatic memory management and improved safety, the senior developer not only helps correct immediate issues but also aids in the junior developer's growth and understanding of modern C++ practices. Additionally, code reviews offer a unique opportunity for reviewers to deepen their understanding of specific features within the project. As they evaluate the work of their peers, reviewers gain insights into new functionalities and complex areas of the application. This enhanced understanding equips them with the knowledge necessary to effectively address future bugs or implement enhancements in those specific areas. Essentially, through the process of reviewing others' code, reviewers not only contribute to the immediate improvement of the project but also prepare themselves to maintain and expand on the project's capabilities in the future.

Mutual responsibility is a key benefit of conducting regular code reviews within a team. As team members consistently examine each other's work, they cultivate a strong sense of shared responsibility and accountability. This collective oversight encourages every member to maintain high standards and thoroughness in their coding efforts. For example, the awareness that their peers will scrutinize their code motivates developers to initially write cleaner and more efficient code. This proactive approach to coding quality reduces the likelihood of future extensive rewrites, streamlining the development process and enhancing overall productivity.

Code reviews frequently catalyze discussions that yield more efficient, elegant, or simpler solutions than those initially implemented. This aspect of code reviews is particularly valuable as it taps into the collective expertise and experience of the team to enhance the overall software design. For instance, consider a developer who implements a function to sort a vector inefficiently. During the code review, another team member might notice the inefficiency and suggest a more effective sorting algorithm or recommend leveraging existing utilities from the standard library. Such suggestions not only improve performance but also simplify the code, reducing complexity and potential for errors, thus making the software more robust and maintainable.

Preparing for code reviews

Before delving into the collaborative process of code reviews, teams must prepare adequately to ensure that these reviews are as effective and efficient as possible. This preparation not only sets the stage for a smoother review process but also minimizes the time spent on avoidable issues, allowing teams to focus on more substantive and impactful discussions.

Clear guidelines

The foundation of an effective code review process is the establishment and documentation of clear, specific coding guidelines tailored to C++. These guidelines should cover various aspects of coding, including style, practices, and the use of language-specific features. By setting these standards, teams create a common language that reduces ambiguity and ensures consistency across the codebase.

Incorporating as much automation as possible into these guidelines can significantly streamline the review process. Tools such as formatters ensure a consistent coding style, while static analysis tools can automatically detect potential errors or code smells before human reviewers ever look at the code. Additionally, ensuring that all code submissions are accompanied by passing unit tests and, where applicable, end-to-end tests can prevent many common software defects from ever reaching the review stage. This level of automation not only saves time for reviewers but also reduces the potential for disputes over subjective preferences in coding style or minor oversights.

Self-review

Another crucial aspect of preparing for code reviews is the practice of self-review. Before submitting their code for peer review, developers should thoroughly examine their own work. This is where tools such as linters and static analysis come into play, helping to catch common issues that can be easily overlooked.

Self-review encourages developers to take responsibility for their code's initial quality, reducing the burden on peer reviewers and fostering a culture of accountability. It also allows developers to reflect on their work and consider potential improvements before involving others, which can lead to higher-quality submissions and more productive review sessions. A developer should methodically evaluate their work by asking the following questions before submitting their code for peer review. This reflective practice helps to refine the code, align it more closely with project goals, and prepare the developer for any subsequent discussions during the code review process:

1. **Did I need to write code? (Is my change rational?)** Before adding new code, consider whether the change is necessary. Evaluate whether the functionality is essential and justify the addition, keeping in mind the potential for increasing complexity in the codebase.

2. **What could I do to avoid writing code? (Is there a third-party library or tool that I can utilize?)** Always look for opportunities to leverage existing solutions. Using well-tested third-party libraries or tools can often achieve the desired functionality without adding new code, thereby reducing potential bugs and maintenance overhead.

3. **Is my code readable?** Assess the clarity of your code. Good code should be self-explanatory to other engineers who might not be familiar with it. Use meaningful variable names, maintain a clean structure, and include comments where necessary to explain complex logic.

4. **Do other engineers need to understand the logic of my code?** Consider whether your code can be understood independently by other developers. It is crucial that others can follow the logic without needing extensive explanations, which facilitates easier maintenance and integration.

5. **Does my code look similar to the rest of the codebase?** Ensure that your code adheres to the established coding styles and patterns in the project. Consistency across the codebase helps maintain uniformity, making the software easier to read and less prone to errors during integration.

6. **Is my code efficient?** Evaluate the efficiency of your code. It's important to consider resource usage, such as CPU time and memory, especially in performance-critical applications. Review your algorithms and data structures to ensure they are optimal for the task.

7. **Do my tests cover edge cases?** Confirm that your tests are comprehensive, particularly checking that they cover edge cases. Robust tests are crucial for ensuring the resilience and reliability of your code under unusual or unexpected conditions.

By answering these questions during the self-review process, developers not only enhance the quality of the code they write but also streamline the peer review process. This thoughtful approach minimizes the likelihood of significant revisions during peer reviews and increases the overall effectiveness of the code review cycle. This preparation can lead to more informed and constructive discussions during code reviews, as developers are already aware of and have addressed many potential concerns.

How to pass a code review

Successfully navigating the code review process is crucial for maintaining the quality of the software and fostering a positive and productive team environment. Here, we outline essential strategies for developers aiming to ensure their code passes a review smoothly and effectively.

Discuss big features with reviewers and code owners before writing code

Before embarking on the development of significant features, it is advisable to consult with code reviewers and owners. This preliminary discussion should focus on the proposed design, implementation approach, and how the feature fits into the existing codebase. Engaging in this dialogue early helps align expectations, reduces the likelihood of significant revisions later, and ensures that the feature integrates seamlessly with other parts of the project.

Go over your code before publishing it

Perform a thorough review of your own code before submitting it for peer review. This self-assessment should cover the logic, style, and adherence to the project's coding standards. Look for any areas of improvement or potential simplification. Ensuring that your submission is as refined as possible not only facilitates a smoother review process but also demonstrates your diligence and respect for the reviewers' time.

Make sure the code is compliant with the code convention

Adherence to established code conventions is critical. These conventions, which cover everything from naming schemes to layout and procedural practices, ensure consistency throughout the codebase. Consistency leads to code that is easier to read, understand, and maintain. Before submitting for review, check that your code strictly follows these guidelines to avoid any unnecessary back-and-forth during the review process.

Code review is a conversation, not an order

It is essential to approach code review as a dialogue rather than a directive. Reviewers typically provide comments and suggestions that are meant to initiate discussion rather than act as unilateral commands. Especially for junior developers, it is important to understand that you are encouraged to engage in these discussions. If a suggestion or correction is unclear, seek clarification rather than silently making changes. This interaction not only aids your professional development but also enhances the collaborative spirit of the review process.

Remember – your code is not you

A vital lesson I learned from my former boss, Vladi Lyga, is that *"Your code is not you."* Developers often invest significant effort and pride in their code, and it can be challenging to receive criticism. However, it's crucial to remember that feedback on your code is not a personal critique of you as a developer or individual. The aim is to improve the project and ensure the highest quality outcome, and this sometimes requires constructive criticism. Detaching their personal identity from their work allows developers to approach feedback more objectively and use it as a growth opportunity.

By preparing thoroughly, engaging in open dialogue, and embracing a constructive perspective towards feedback, developers can effectively navigate the code review process. These practices not only enhance the quality of the code but also contribute to a more supportive and collaborative team environment.

How to efficiently dispute during a code review

Disagreements are a natural part of the code review process. Different perspectives can lead to conflicts over approaches, implementations, or interpretations of best practices. Handling these disputes efficiently is crucial to maintaining a productive review process and a healthy team environment. Here are key strategies to manage disagreements effectively during code reviews.

Clear justification for changes

It is imperative for reviewers to not only point out areas needing improvement but also to clearly explain why changes are necessary. When suggesting a modification, a reviewer should provide a rationale that is anchored in best practices, performance considerations, or design principles relevant to the project. Including links to coding standards, articles, documentation, or other authoritative resources can greatly enhance the clarity and persuasiveness of the arguments. This approach helps the reviewee understand the reasoning behind the feedback, making it more likely for them to see the value of the suggested changes.

Reciprocal explanation from reviewees

Similarly, if a reviewee disagrees with a comment or suggestion, they should also articulate their reasoning clearly. This explanation should detail why their approach or solution was chosen, supported by relevant technical justifications, documentation, or precedents within the project. By providing a well-reasoned argument, the reviewee fosters a more informed discussion, which can lead to a better understanding or an improved solution.

Direct communication

When a dispute involves more than a couple of comments back and forth, it's advisable to shift the conversation from written comments to a direct dialogue. This can be through a video call, phone call, or face-to-face meeting, depending on what's feasible. Direct communication can prevent the kind of miscommunication and escalation often seen in text-based discussions, which can quickly become unproductive and contentious, much like lengthy threads seen on platforms such as Reddit.

Involving additional perspectives

If a resolution cannot be reached between the reviewer and the reviewee, it can be beneficial to involve additional perspectives. Bringing in a third engineer, a product manager, a QA specialist, or even an architect can provide new insights and help mediate the disagreement. These parties may offer alternative solutions, compromise approaches, or a decision based on broader project priorities and impacts. Their input can be crucial in breaking deadlocks and ensuring that the decision is well rounded and aligns with overall project goals.

Efficient dispute resolution during code reviews is vital for keeping the review process constructive and focused on enhancing the quality of the codebase. By explaining the rationale behind feedback, encouraging direct communication, and involving additional perspectives when necessary, teams can resolve disagreements effectively and maintain a positive, collaborative environment. This approach not only resolves conflicts but also strengthens the team's ability to work through future challenges collaboratively.

How to be a good reviewer

The role of a reviewer in the code review process is crucial not just for ensuring the technical quality of code but also for maintaining a constructive, respectful, and educational environment. The following are some key practices that define a good reviewer.

Initiate the conversation

Start the review process by initiating a friendly conversation with the reviewee. This can be a brief message acknowledging the effort they have put into the **pull request** (**PR**) and setting a positive tone for the upcoming review. A cordial start helps to build rapport with the reviewee, making the subsequent exchange more open and collaborative.

Maintain politeness and respect

Always be polite and respectful in your comments. Remember that the reviewee has invested significant effort into their code. Critiques should be constructive, focusing on the code and its improvement rather than on the individual. Phrasing feedback in the form of questions or suggestions rather than directives can also help to keep the tone positive and encouraging.

Review manageable chunks

Limit the amount of code you review at one time to about 400 lines, if possible. Reviewing large chunks of code can lead to fatigue, which increases the likelihood of missing both minor and critical issues. Breaking down reviews into manageable parts not only enhances the effectiveness of the review process but also helps to maintain a high level of attention to detail.

Avoid personal bias

While reviewing, it's important to distinguish between code that must be changed for objective reasons—such as syntax errors, logical errors, or deviations from project standards—and changes that reflect personal coding preferences. For example, let us consider the following code snippet:

```cpp
std::string toString(bool done) {
  if (done) {
    return "done";
  } else {
    return "not done";
  }
}
```

Some engineers might prefer to rewrite this function the following way:

```cpp
std::string toString(bool done) {
  if (done) {
    return "done";
  }
  return "not done";
}
```

While the second version is a bit more concise, the first version is equally valid and adheres to the project's coding standards. If you feel strongly about a personal preference that could enhance the code, clearly label it as such. Indicate that it is a suggestion based on personal preference and not a mandatory change. This clarity helps the reviewee understand which changes are essential for compliance with project standards and which are optional enhancements.

Focus on understandability

One of the most critical questions to ask yourself as a reviewer is whether the code is understandable enough that you, or someone else on the team, could fix it in the middle of the night. This question cuts to the heart of code maintainability. If the answer is no, it's important to discuss ways to improve the code's clarity and simplicity. Code that is easily understood is easier to maintain and debug, which is crucial for long-term project health.

Being a good reviewer involves much more than just identifying flaws in code. It requires initiating and maintaining a supportive dialogue, respecting and acknowledging the efforts of your peers, managing your review workload effectively, and providing clear, helpful feedback that prioritizes the project's standards over personal preferences. By fostering a positive and productive review environment, you contribute not only to the quality of the code but also to the growth and cohesion of your development team.

Summary

This chapter delved into the essential practices and principles of conducting effective code reviews, a critical component of the software development process in C++. Through a series of structured sub-chapters, we have explored various aspects of the code review process that collectively ensure high-quality, maintainable code while fostering a positive team environment.

We began by discussing the *origins of code reviews*, introduced in the 1970s by Michael Fagan, highlighting its transformational role in moving software development from an isolated task to a collaborative effort that enhances code quality and reduces bugs.

In the *Preparing for code reviews* section, we emphasized the importance of clear guidelines and self-review. Developers are encouraged to use tools such as linters and static analyzers to refine their code before it undergoes peer review, ensuring adherence to coding standards and reducing the iterative cycle of code revisions.

The *How to pass a code review* section outlined strategies for developers to ensure their code is well received during reviews. This includes discussing significant changes before coding, understanding code reviews as a constructive dialogue, and remembering to detach personal identity from the code to view feedback objectively.

The *How to efficiently dispute during a code review* section addressed handling disagreements productively. We discussed the importance of clear justifications for changes, using direct communication to avoid misinterpretations, and involving additional perspectives when necessary to resolve conflicts and reach a consensus.

Finally, in the *How to be a good reviewer* section, we provided guidance on initiating reviews with a positive interaction, reviewing code in manageable portions, avoiding bias from personal preferences, and assessing code for its clarity and ease of understanding in critical situations.

Throughout this chapter, the underlying theme has been that code reviews are not just about critiquing code but about building a supportive community of developers who share knowledge, improve continuously, and aim for excellence in their coding practices. The goal is to enhance both the technical quality of the software and the professional development of the team members involved. By adhering to these best practices, teams can achieve more robust, efficient, and error-free code, contributing significantly to the success of their C++ projects.

Index

packtpub.com

Subscribe to our online digital library for full access to over 7,000 books and videos, as well as industry leading tools to help you plan your personal development and advance your career. For more information, please visit our website.

Why subscribe?

- Spend less time learning and more time coding with practical eBooks and Videos from over 4,000 industry professionals

- Improve your learning with Skill Plans built especially for you

- Get a free eBook or video every month

- Fully searchable for easy access to vital information

- Copy and paste, print, and bookmark content

Did you know that Packt offers eBook versions of every book published, with PDF and ePub files available? You can upgrade to the eBook version at packtpub.com and as a print book customer, you are entitled to a discount on the eBook copy. Get in touch with us at customercare@packtpub.com for more details.

At www.packtpub.com, you can also read a collection of free technical articles, sign up for a range of free newsletters, and receive exclusive discounts and offers on Packt books and eBooks.

Other Books You May Enjoy

If you enjoyed this book, you may be interested in these other books by Packt:

C++20 STL Cookbook

Bill Weinman

ISBN: 978-1-80324-871-4

- Understand the new language features and the problems they can solve
- Implement generic features of the STL with practical examples
- Understand standard support classes for concurrency and synchronization
- Perform efficient memory management using the STL
- Implement seamless formatting using std::format
- Work with strings the STL way instead of handcrafting C-style code

Hands-On Design Patterns with C++ (Second Edition)

Fedor G. Pikus

ISBN: 978-1-80461-155-5

- Recognize the most common design patterns used in C++
- Understand how to use C++ generic programming to solve common design problems
- Explore the most powerful C++ idioms, their strengths, and their drawbacks
- Rediscover how to use popular C++ idioms with generic programming
- Discover new patterns and idioms made possible by language features of C++17 and C++20
- Understand the impact of design patterns on the program's performance

Packt is searching for authors like you

If you're interested in becoming an author for Packt, please visit `authors.packtpub.com` and apply today. We have worked with thousands of developers and tech professionals, just like you, to help them share their insight with the global tech community. You can make a general application, apply for a specific hot topic that we are recruiting an author for, or submit your own idea.

Share Your Thoughts

Now you've finished *Refactoring with C++*, we'd love to hear your thoughts! Scan the QR code below to go straight to the Amazon review page for this book and share your feedback or leave a review on the site that you purchased it from.

`https://packt.link/r/1837633770`

Your review is important to us and the tech community and will help us make sure we're delivering excellent quality content.

Download a free PDF copy of this book

Thanks for purchasing this book!

Do you like to read on the go but are unable to carry your print books everywhere?

Is your eBook purchase not compatible with the device of your choice?

Don't worry, now with every Packt book you get a DRM-free PDF version of that book at no cost.

Read anywhere, any place, on any device. Search, copy, and paste code from your favorite technical books directly into your application.

The perks don't stop there, you can get exclusive access to discounts, newsletters, and great free content in your inbox daily

Follow these simple steps to get the benefits:

1. Scan the QR code or visit the link below

https://packt.link/free-ebook/9781837633777

2. Submit your proof of purchase
3. That's it! We'll send your free PDF and other benefits to your email directly

www.ingramcontent.com/pod-product-compliance
Lightning Source LLC
Chambersburg PA
CBHW080614060326
40690CB00021B/4692